COMPETITION CAR
DATA LOGGING

COMPETITION CAR
DATA LOGGING
A PRACTICAL HANDBOOK

Simon McBeath

Haynes Publishing

© Simon McBeath 2002

All rights reserved. No part of this publication may be reproduced, stored in a retrieval system or transmitted, in any form or by any means, electronic, mechanical, photocopying, recording or otherwise, without prior permission in writing from the publisher.

First published in 2002
Reprinted 2003

A catalogue record for this book is available from the British Library

ISBN 1 85960 653 9

Library of Congress catalog card no. 2001135406

Published by Haynes Publishing, Sparkford,
Yeovil, Somerset BA22 7JJ, UK

Tel: 01963 442030 Fax: 01963 440001
Int. tel: +44 1963 442030 Fax: +44 1963 440001
E-mail: sales@haynes-manuals.co.uk
Web site: www.haynes.co.uk

Haynes North America, Inc.,
861 Lawrence Drive, Newbury Park,
California 91320, USA

Printed and bound in England by J. H. Haynes & Co. Ltd, Sparkford

Jurisdictions which have strict emission control laws may consider any modifications to a vehicle to be an infringement of those laws. You are advised to check with the appropriate body or authority whether your proposed modification complies fully with the law. The author and publishers accept no liability in this regard.

While every effort is taken to ensure the accuracy of the information given in this book, no liability can be accepted by the author or publishers for any loss, damage or injury caused by misuse of, errors in, or omissions from the information given.

Author's note
Although the abbreviation for gravitational acceleration is sometimes a lower case g, because of its importance in this book and for the sake of clarity, it is expressed with a capital G.

Contents

	Author's preface	6
	Acknowledgements	7
Chapter 1	**The basics**	8
Chapter 2	**DAS hardware**	17
Chapter 3	**Basic analysis tools**	26
Chapter 4	**Rpm analysis**	36
Chapter 5	**More channels!**	54
Chapter 6	**Logging speed**	68
Chapter 7	**On the throttle**	77
Chapter 8	**Steering angle**	85
Chapter 9	**Lateral G**	93
Chapter 10	**Longitudinal G**	102
Chapter 11	**Software extras**	111
Chapter 12	**Yet more channels?**	120
Chapter 13	**Data logging in use**	130
Appendix A	**First unpack the computer!**	147
Appendix B	**Mini supplier directory**	153
	Glossary of terms and abbreviations	156
	Index	159

Author's preface

IT WAS SOMEWHAT startling to discover that much of the data acquisition capability that we're going to look at in this book has been around, in principle, if not technologically, since the 1960s. The big difference between then and now is that the ability to log data has become so much more accessible. The low cost of personal computers and other electronic technology has made this possible. It is now within many, if not most competitors' budgets to obtain some form of data logging equipment and benefit immediately from the extra knowledge that will be gained.

This was one of the most appealing aspects of my own introduction to data logging. The torrent of new information that suddenly appears can lead to very rapid increases in performance. Guesswork becomes a thing of the past, at least in relation to the parameters that you are logging, and solid information provides a much better foundation on which to make judgements. But the learning process never stops. It's true that those early, often substantial gains tend to tail off after a while as you would expect, but the information you collect never loses its value, and often you can be brought up short by 'rediscovering' something quite basic. And from anecdotes I heard during my researches, this principle goes right to the top of the motorsport technology spectrum.

Data logging provides information in three categories:
1. Information to be conveyed to the driver on a dash display, usually restricted to a few basic facts such as rpm and lap time.
2. Vehicle system data, ranging from the simple and cheap (e.g. oil pressure) to the expensive (tyre pressures).
3. Data related to engine, driver and chassis (including aerodynamic) performance, ranging from rpm and speed through to wheel hub accelerations.

It is the latter category on which we concentrate in this book because that's where performance gains principally come from.

But if there is one phrase to go by in data logging it is 'remain open-minded'. Drivers in particular need to not regard data logging as 'the spy in the cab', but instead as an honest friend who tells it like it is. As one race engineer put it to me, some drivers greet data that shows they could improve with 'OK, put the bloody data logger in the seat and see if it can do any better!' The open-minded ones will say: 'Right, let's look at the data and see how I can go quicker.' But race engineers need to be just as open-minded when it comes to data analysis. Just because something hasn't happened before doesn't mean it can't happen now, or later. Take nothing at face value – check everything.

Another very relevant saying is: 'don't over-complicate things'. Data analysis takes time, so keep your system as simple as possible in order that you can maximise the benefits. Even the simplest data logging equipment can provide incredible insight.

Acknowledgements

THIS IS ALWAYS the toughest section of a book to write because you just know you're going to miss out somebody important, somebody who went out of their way to provide help without which the whole project would never have been finished. As always, apologies to anyone in that category who slipped from my mind at the critical moment.

Major thanks must go to Mick Kouros, top race engineer and top bloke, who helped me more than he realises; to Jim Hersey at Astratech, and to Mike Holmes, Iain Wight and George Lendrum at Pi Research.

Thanks too to Phil Abbott and Mick Hyde at Radical Motorsport, Martyn Bell, Steve Crabtree at Stack Ltd, Tom Brown, Brandon Fry, Keith Wiggins and Paige Pell at Herdez Bettenhausen, Nigel Greensall, Kelvin Jouhar, Alan Lis, Fraser MacKellar at MoTeC (Europe), Richard Marshall, Andy McBeath, Tim Mullis at Race Data Solutions. and John Nixon at Cranfield University.

It's always a pleasure and an instructive experience talking to race engineers and fellow competitors, so my thanks to all of those with whom the topic of data logging was casually raised in conversation.

Special thanks as ever to my partner Tracey, whose unwavering support and assistance with research kept me going. The line drawings in this book are hers too as are some of the photos, and her ability to cut the gobbledegook out of computing formed the basis of the first appendix.

Finally, a particular thanks to all at Haynes Publishing for their patience. I got there in the end.

Chapter 1

The basics

IT ALWAYS HELPS to start off with definitions, and a book on data logging is no exception to this general rule. Data logging can be said to take over where ordinary vehicle instruments stop. Whereas rev counters, oil pressure gauges and so forth very helpfully measure and display certain parameters from around the car, that's all they do. If the driver is to make use of them he or she has to take instantaneous notice of the values displayed and take appropriate action. If one of the gauges gives some particularly interesting readings for some reason, then the driver has to remember what the readings were and when they occurred in order to either investigate the cause(s) or pass the information to another team member to do that job. Data logging, or data acquisition to give it its posh name, not only measures parameters around the car, it also records the values so that the results can be studied in detail later on, and a permanent record can also be made for future reference. The useful information that can be gleaned is thus many times that which comes from basic engine instruments.

Close study of data yields masses of useful information!

It is also useful to define too what data logging isn't. It isn't telemetry. Telemetry is often confused with data acquisition, and it's not surprising. But in its literal sense, telemetry means 'measure far away', and it refers to the transmission of data (possibly from a data logging system) by radio or similar means to a remote receiver, which in the context of motorsport usually means from a car to the pits, or perhaps to a team base even further removed than that. So telemetry is an extension to data logging in that sense, but it is not the same thing. Whilst we'll touch on telemetry in the last chapter when we look briefly at how data acquisition is done in the top levels of motorsport, the rest of this book is just about data logging.

Having got that particular bee out of my bonnet, let's go back to the basics again. How do data logging systems work? It was stated above that a data acquisition system (DAS) measures and records the values of anything from one to a multitude of parameters around the car. These are just two aspects to making use of data logging. There are also two further aspects that are important – accessing the recorded data, and examining and analysing it. So there are four elements to a DAS, and putting the conventional terms to these we have; 'sensors' to measure the parameters in which we are interested; the DAS 'unit' itself which contains the electronic memory that stores the data; a connection or 'download cable' to transfer the data; and a personal computer (PC) on which is loaded the 'analysis software' that enables examination and study of the data.

At this point, computer-phobics will be wishing either that they hadn't bought this book, or have just decided that they won't. But please don't do either pf those things! Chapter 3 and Appendix A will offer a novice's guide to computing which will enable you to navigate around the thing so that at least you can use it with a DAS. The main problem with computers is that the people who know how to use them, or who sell them (the two groups do not necessarily overlap) are generally incapable of explaining how to use them in comprehensible terms to people who don't know how to use them. It isn't that difficult, as computer-literate folk always tend to say, and the benefits in knowledge you will gain about your competition car's performance and that of its driver are considerable. So if you don't already, it's definitely worth getting to know your way around a PC.

What kind of data?
Conventional instruments are generally restricted to displaying engine-related parameters, but the data that can be gathered with a DAS falls into three generic categories: engine, chassis and driver. Naturally, these are inter-related in many cases because a competition car is a complex system in which all three of those major components are connected by a set of controls.

Engine-related parameters that can be 'logged' include engine rpm, fuel and oil pressure, water and oil temperature, turbocharger boost pressure, exhaust gas temperature, battery voltage, inlet air temperature, and throttle position.

Chassis-related parameters that may be logged include wheel speed, steering angle, lateral and longitudinal G-force (to indicate cornering and braking forces in particular), brake line pressures, damper movement and gear position. More 'advanced' parameters also include the measurement of ride height, driveshaft or propshaft torque, suspension loads, air speed, tyre pressures, tyre temperatures and brake disc temperatures, and aerodynamic parameters including air speed and local air pressures.

Driver-related parameters are those chassis (and a couple of engine) related parameters over which the driver has direct control via his or her input, such as throttle position, gear, steering angle and brake line pressure, or those which are directly influenced by driver input, such as engine rpm, speed, and longitudinal and lateral G-force.

Clearly some of these are partly down to driver input and partly to the engineering of the chassis, and in those cases, study of the results can be used to find improvement in the performance of both. Arguably the minimum list of key parameters worth studying in order to assess driver and chassis performance can be whittled down to just speed (with or without rpm), throttle position and steering angle. After all, when all is said and done, speed is what it's all about, and the throttle is what gets the car up to speed, whilst the combination of steering angle and throttle angle can tell you a lot about a car's handling.

Many would also add lateral G-force to that short list too, and for one good reason – most DASs log lateral G so that the analysis software can use the data, in conjunction with speed, to compute and display a 'virtual track map'. This is a map which actually shows the path taken by a car during a run or a lap of a track, and it is displayed by the DAS software on the computer screen as a line graphic that looks just like what it says it is – a map of the track. A virtual track map is a useful tool when studying logged data because the software shows graphically where the car is on the track, usually with a little dot or a cross, and also where in the other data plots this position is, marked by a vertical line or 'cursor'. In this way it is easy to mentally connect the data graphs to the position of the car on the track, and read off, for example, at any one point on the track what rpm, speed, lateral G, steering angle and so on were occurring at that instant.

The track map is a convenience though rather than an essential item, especially if you drive on simple or shorter courses like hillclimbs and less-convoluted circuit layouts – here it is possible to quickly develop the ability to mentally overlay the graphs on to the kind of course map you find printed in event programmes. On more complex circuits however, virtual track maps are a useful add-on.

However, an awful lot of very useful and revealing data can be derived from the four main parameters, rpm, speed, throttle angle and steering angle, and curious though it may seem, you can also learn quite a large proportion of that from just looking at a trace of engine rpm. Recording engine rpm is the simplest form a data logging. As well as the ignition system's electrical impulses driving the rev counter display, they are also counted and recorded in an electronic memory. This is really where useful data logging began, and the recording or 'intelligent tachometer' was the instrument that provided the capability, and is still the cheapest way to get into basic data logging. We'll delve into the details more deeply in Chapter 4, but rpm doesn't just tell you what your engine history is, it tells you whether your gearing is right, whether your engine's state of tune is suitable for the events you are doing, what your speed is at all times, and it can even provide pointers and reminders about driver and chassis behaviour.

Sensors and channels

To measure and log other parameters requires that specific sensors must be installed on the car. Sensors come in various types, shapes and sizes, and we'll look at these in more detail in Chapter 2. But whatever type of sensor is being discussed, its purpose is to provide a measurable electronic signal that can be recorded by the DAS. Thus, for example, it has to measure how far open the throttle is and convert that measurement into a signal that the DAS can detect and log. So, a suitable sensor could be one that measures linear movement, if the measurement is made at the pedal, or at the engine end of the throttle cable or on a throttle slide arrangement, or perhaps one that measures rotational movement if the measurement is to be made on the throttle spindle on butterfly or barrel throttles.

Specific sensors can measure properties like temperatures, pressures, accelerations, stresses and strains in materials, and changing magnetic fields, and convert the measurements into electronic messages that can be sent to a DAS. The DAS then compares the signals to its stored calibrations, and logs the values (of throttle angle, water temperature or whatever) in its memory.

Each sensor connected to a DAS constitutes a 'channel'. DASs are all limited as to the maximum number of channels they can record, whether they are single-channel rpm only devices like recording tachometers, or a full professional system that might offer over 100 different channels. The point to keep in mind is that each channel can only log one parameter at a time, so if you have a four-channel logger, you can record four parameters at once. You could have more than four sensors on the car though, and perhaps log certain channels in practice but swap over to different ones for actual competition, and this is one way to make good use of a low-cost DAS if you are on a tight budget (which of us isn't?).

Sampling rates and memory

Browsing through the suppliers' literature you'll almost certainly come across statements referring to sampling rates, or sampling frequencies. These refer to the number of times per second that data is recorded. At the low-cost end of the data logging market, sampling frequency may be 10Hz, or 10 samples per second, whist intermediate systems might typically offer double that rate, 20Hz. More advanced systems not only offer the ability to log much more rapidly if required (up to 1,000Hz), but can also allow different channels, selected by the user, to be logged at different rates, and can even be set to log at very slow sampling rates too (maybe 1Hz or slower) on parameters where rapid sampling isn't necessary, for example water temperature, which could happily be measured once every 10 seconds in most instances, never mind 10 times a second.

Some parameters might need to be logged at high frequency though if important and possibly crucial data is not to be missed – for example, damper piston movement can be very rapid on some competition cars, up to and even exceeding 300mm per second (12in per second) over a kerb or a rock. A sampling rate of 10 Hz, where samples are taken every 0.1 seconds, could easily miss out more rapid and possibly vital elements of the suspension movement (a damper piston could move 15mm, or 0.6 inches, at 300mm per second in 0.05 seconds, and the movement would be completely missed by a DAS sampling every 0.1sec). Then again, it depends on why you want to measure damper movement as to whether fast sampling is important – if you were only interested in the slower movements of the car's suspension, such as pitch, roll and dynamic changes in ride height due to aerodynamic loads, slower rates might actually filter out unwanted data by just not recording it.

The number of channels logged and the sampling rate combine to determine how much memory is required in the DAS to store data over a given time period. Simply, more channels sampled more frequently creates more data to be stored in the same time period. This could be put the other way around too, in that the amount of memory in the DAS dictates how many channels and what logging frequency are going to be practicable over the duration of events (test or practice sessions and actual competitions) that you want to record. Systems with more memory cost more money, making this another budgetary consideration. So you either need to temper your enthusiasm to log every parameter you can at high sampling rates to something that is actually possible on the budget available, or else you'll have to raise some more cash! The DAS suppliers will be happy to guide you, and reputable systems are 'modular', allowing you to add more channels, or more memory as and when you can.

Different motorsport disciplines also create different demands for memory too. A drag racer might only want to record between five and 15 seconds worth of track time at any one go before being able to download the data to a computer for analysis; a hillclimber might want to store four or five runs (or even just one run) of 30 to 60 seconds; a club circuit racer or a stage rally driver would need to be able to store chunks of around 10 to 20 minutes; a Formula 1 team might need to be able to store a couple of hours of data, and so on. You can now begin to see that those in the upper echelons of motorsport, who sample the most channels, at the highest frequencies, and for the longest periods, not only create many times more data for study and analysis than those of us in club and national level disciplines, but they also have to spend a lot more money in order to log, store and analyse it!

Downloading and analysis

Downloading data is the process of collecting the information stored in the DAS on-board memory and transferring it to a computer. This is done by connecting the computer (usually a portable laptop, but not always) to the DAS with the cable supplied with the system, and then entering the appropriate magic words or commands via the computer keyboard or its mouse or pointing device (don't worry, if this sounds like gibberish, there will be more on computer terminology and operation principles in Chapter 3 and Appendix A), and the DAS software.

The electronically stored data then passes along the cable from the DAS to the PC and into the PC's own memory, ready for analysis and more permanent storage. The data is stored in a PC's memory in 'files', analogous to the paperwork variety, to each of which you give a unique name that will enable you to track it down easily later, such as 'SILVP401', meaning Silverstone practice, April 2001.

Downloading data from the on-board data acquisition system.

More on this later too, but it is important that your PC files are well organised, just like any other filing system that will eventually contain a lot of information.

Once transferred to the PC the logged data can then be viewed and examined in various ways. You can, if you wish, look at tables of figures, but the best way, which usefully and not coincidentally is also the way in which the DAS software packages are designed to show the information, is to look at it in the form of graphs and charts. It is said that a picture can be worth a thousand words. If that is so, then a graph is worth at least as many numbers in a table! Data is often shown in graphs of the logged parameter, be that engine rpm, speed or whatever, plotted against time or distance. This means that the rpm (speed etc) is shown on the vertical, or y-axis of a graph, and time or distance is shown on the horizontal, or x-axis of the graph. Thus, the wiggly line, or 'trace' across the body of the graph shows how the particular parameter you are looking at varies with time or distance.

The data trace starts when the DAS commences recording, which is either when it is powered up, or when an appropriate button is pushed to initiate recording. In some cases recording starts automatically when a wheel speed sensor detects forward motion, which can be a useful way of only recording data of interest, because the DAS will ignore periods like engine warm ups when the car doesn't actually go anywhere. Recording then stops, usually when the system is powered down again, which means the last thing in the DAS's memory is the last thing the car did before powerdown. This means that the data graphs will fill the latter part of the system's memory, and in order to find the bits you are interested in, you use your PC to 'scroll' through the graphs, like using the fast rewind on a tape recorder, until you come to the parts that you want to study.

It might at first appear as though the collection of wiggly traces on the PC screen is without form or shape, but pretty soon after starting to examine them you begin to see patterns emerging which represent what the driver and the car have been up to. For example, taking the simplest system, an rpm recorder, you will first of all be able to pick out the interesting bits of the trace, as against the dull bits where you were just warming the engine and driving to the start line or grid, because the rpm peaks will be more pronounced in the interesting bit – well anyway, they should be! Then you will pick out the characteristic rising and falling skewed saw-tooth trace that represents going up through the gears. Then the rpm troughs that represent the slowest parts of each corner will become evident, and soon you will recognise the pattern that depicts an entire run or lap of the track data your are studying, and in doing this you are mentally overlaying a map of the track on to the data by spotting the parts of the trace that represent particular positions of the track. Of course, if you have a DAS that shows a track map on-screen, part of this is done for you, but the ability to relate the graph traces directly to track position will nevertheless unconsciously develop as you become familiar with the data for a given track.

The same goes for the other logged channels that measure the car's performance in some way, although the fit of something like an engine temperature trace to a track layout might not be so obvious, nor would it necessarily be particularly worthy of study unless for some reason your car was marginal on cooling . . . The fit of the traces of real interest does once again become obvious pretty soon after you start looking at the data. However, it is in the detail of the traces that the real interest lies, and where the genuine benefits that can be gained from the use of data logging are to be found. Again, using rpm only for example, comparing the rpm through the same corner on different runs or laps will tell you if there were speed differences in that corner on the different occasions. There may be a variety of reasons for any differences, and the differences may only be small, but the data is there, in front of you on-screen, giving you hard, objective and irrefutable evidence of driver and car performance. Some drivers have expressed alarm at this 'spy in the cab'. Others say they don't need a data logging system to know when they've done something wrong. Both groups are missing out on the chance to learn and improve their driving and their car.

How the data is to be analysed and interpreted we shall go into in much greater detail in later chapters. But the diligent analysis of the logged data will yield much information about driver and chassis performance, not forgetting the engine of course. This is what data logging is all about.

Types of data logging systems

Data logging has moved on a bit from the stopwatch, notebook and pen, and there are now systems available to suit most budgets, starting with the recording tachometers at just a few hundred pounds (middle hundreds of US dollars) including analysis software, to a simple four-channel system for a few more hundreds of pounds (just into four figures in US$). Around a couple of thousand pounds (3,000 US dollars or so) will get you a decent 10-channel system, and for a fully featured 127-channel professional system, well you'll just have to put your house on the market.

But before you even consider how much money you want to spend on a data logging system, it's also worth considering that analysing the data takes up a lot of time, both at events and in between them, and there is never enough time to fully analyse the data, even from a single-channel rpm recorder at events. This means it may well be pointless buying yourself a top-of-the-range 127-channel DAS if you compete at weekends and then spend your weekdays in gainful occupation in order to pay for your competition. The evenings in between are not going to be anywhere near enough time to sift each weekend's data mountain. And besides, you've probably got to prepare the car, and your friends and family might like to see you once in a while too.

On the other hand, you may be in a position where you can spend time in the week between events, or maybe hire the time of a 'data acquisition guy', or a 'DAG', who can do that for you, in which case you will be able to make use of more data. So cut your cloth according to your needs, your ability to exploit your DAS, and your budget.

Let's now take a brief, non-commercial tour around the market place to see what types of systems are available. As above, I'm going to resist the temptation to mention actual prices because these things change rapidly in the context of a book's life, although rough guidance is given. However, you will find a list of DAS suppliers in Appendix A at the back of the book, and a few phone calls or website visits will determine product ranges and up-to-date prices. From here onwards, the mention of specific company names should not be taken as a particular endorsement of that company's products, neither should omission of any company name be regarded in any way as a negative opinion. I simply have more knowledge of some companies' products, either through having personal experience with them in competition myself, or through specific assistance for this book from the companies concerned.

Recording tachometers

The recording tachometer, or intelligent rev counter, can be regarded as an 'entry level' DAS. Oxfordshire, UK, company Stack were the first on the scene here, supplementing their steady needle display, stepper motor-driven analogue rev counters with initially a few minutes of memory, but at the time of writing, offering 25 minutes of memory as standard. The cheapest model allows play back of earlier running on the tacho dial, which precludes the need for a link with a computer and whilst this may appeal, if only for the cost saving over the more advanced model, it is only a little more use than the old-fashioned rev counter tell tale, and nothing like as useful as being able to study the data on-screen, or in the form of a printout, in your own time.

To be regarded as a useful data logging tool then, the recording tachometer really does need to be supplemented with the ability to link up with a computer, and whilst the requisite leads, connections and computer software currently adds 50 per cent to the cost of the tacho, the overall cost is still less than a cheap set of racing tyres, and its benefits will not wear out after a few events. This assumes you do not need to buy a computer, but since that cost has to be borne to make use of any of the systems discussed in this book, the tyre-cost comparison is valid. We'll look in detail at what can be learned from a simple rev trace in Chapter 4.

Intermediate systems

For a little over double the cost of the recording rev counter and the paraphernalia needed to connect to a PC, a four-channel DAS with sensors, leads, connections and software is available from another UK company, Astratech Racing Technology. The basic kit logs wheel speed and engine rpm plus two other user-selected channels, which are usually throttle position and lateral G-force, the latter enabling the software to compute and display virtual track maps. This kit also comes with an infrared lap timing beacon and an LCD display to feed the driver information during running. Other sensors are available, such as steering angle, water and oil temperatures, oil and fuel pressures and so forth, so the choice for the two user-selectable channels is fairly wide. You can add more channels later.

The Astratech system logs at 10Hz, so it is one of the slower systems on the market, but it is also the cheapest available in the UK at least. It also offers very simple-to-use software (you will often read the word 'intuitive' in connection with easily used software, but don't believe it for a second unless you are already a computer geek who speaks the language of computer users!), which uses animated graphic images ('icons') of a foot on a throttle, or a steering wheel or whatever to supplement and aid interpretation of the graph traces. If you're not familiar with the use of graphs, this may prove very helpful, especially in the early stages of data analysis, and even if you are fully conversant with graph interpretation, the animated icons help to point out areas of particular interest very rapidly, which can then be studied more closely by examining the graph traces.

Next up come the more costly data loggers from companies such as Stack, Motec and Pi Research. For an investment amounting to between three and five times the cost of the recording rev counter, or around £1–2,000 (maybe $1,600–$3,000 or so) you move into real multi-channel territory, with anything from six to 20 channels, and also more complex and sophisticated, but not necessarily such easy-to-use software. The better systems allow you to upgrade as you need, adding sensors to spare channels, updating the software as new improved versions become available, and even adding logging memory.

For the bulk of this book we will restrict ourselves to going no further than this level of 'advanced intermediate' data logging system, although clearly you can spend as much as the cost of a club competition car and more on a DAS if you feel the urge. We will look at the top professional systems out of interest later, but there is plenty to be learned from a few well chosen channels, and certainly enough to keep most competitors glued to their PC screens for nights on end between events!

PC literacy

It has been stated that personal computers are an integral part of the analysis of

Data logging adds to driver feedback, but doesn't replace it.

logged data, and without a computer there is very little useful analysis that can be done. Clearly then, in order to get to grips with data acquisition analysis you need to be 'computer literate' at least to the point where you can find your way around the software provided. But don't worry, you do not need to be a computer programmer to use the software the DAS suppliers offer with their products, and in any case if you are a complete computer novice, Appendix A will take you through the early learning stage. But you do need to be willing to take yourself to a basic level of computer literacy to be able to benefit from data logging.

What you get and what you don't
Data acquisition provides you with objective evidence, from whatever number of channels you log, of what has been happening to those parameters during the running of your competition car. It does not explain how that data occurred or why, but with thoughtful analysis it can provide some pointers to explain what the engine, chassis and driver have been doing. Combined with some knowledge of 'racecar engineering' (the word 'racecar' should be regarded as generic here) and a healthy dose of common sense, these pointers can then be used to define strategies for improvement to the engine, chassis, driver or any combination of these three. Comparisons of performance 'before' and 'after' modifications are made can be achieved objectively, not just on the basis of overall lap or run time but in specific areas of vehicle or driver performance.

There are two things to keep in mind however. First, none of this data can replace the driver's feedback and subjective assessment of a car, because the driver has to feel at ease in a car to make it go consistently quickly. Data simply adds to the driver's feelings and recollections. Secondly, to repeat something I've already stated, don't be misled into thinking that you need to spend a year's salary on a DAS to gain useful information. You can get a whole lot of useful facts and figures from the simple and intermediate systems we'll concentrate on in this book.

Chapter 2

DAS hardware

DATA ACQUISITION SYSTEM hardware consists, as we have seen, of sensors, a logging unit, and all the wiring and connections to enable the sensors to communicate with the logger, and the logger to communicate with your computer (that bit we'll look at in the next chapter). However, in this chapter we're going to take a look at the generalisations of DAS installation, how to protect your system from the harsh environment within a competition car, and how the various sensor types do their job.

DAS installation necessarily includes the mounting of sensors, the logger units themselves, the cables and connectors and any bracketry required for those purposes. Specific details will vary from system to system, and if you are not procuring the services of your supplier or other qualified installer to fit your DAS then it's a matter of following the supplier's fitting instructions as closely as you can. No supplier of any product can make all things for all applications so a degree of improvisation may be needed. But all the people at the DAS suppliers that I have had the opportunity to speak to have been ready and willing to help with advice and hands-on assistance if required, which is only fair if you're spending big chunks of money on their products.

Component location

The DAS unit itself is best located in the cockpit or cabin of your competition car where it obtains the best protection from the external environmental of the weather, and the internal effects of heat from the engine and associated components, not forgetting spurious signals from other electrics such as the ignition system.

Sensors need to go wherever a measurement is to be made, an obvious statement unless you've been watching too many episodes of *Star Trek* and you actually believe that sensors can pick up readings from light years away! Some types of sensors do not actually make contact with the object they are measuring, but they are still likely to be in close proximity. Thus sensors will be located around the engine, on wheel hubs, on the steering column and so forth as required by their function. Accelerometers, or G-sensors as they are more usually referred to, may be integrated into the logger unit or they may be separate from it, but in either case they need to be mounted close to the car's centre of gravity. Sensors are expensive and fragile components and need to be treated as such if they are to work reliably and durably.

The cables that connect the sensors to the data logger need to be thoughtfully routed and fastened. Apart from making a tidy job of their installation, you need to follow the manufacturer's instructions on isolating sensor cables from other electrical and electronic systems or cables to avoid picking up stray signals that could interfere with the reliable collection of data signals. Pay attention too to sources of extreme heat and avoid mechanical contact with moving parts, throttle linkages and so forth. Make sure that cables cannot be damaged during routine or non-routine maintenance or disassembly, be it removal of the engine or simply twirling a socket through a gap in the chassis. Never put cables under any tension because vibration will then do its best to disconnect them.

Use the manufacturer's supplied connectors, and fit them in the manner described – that way you can shout at him if they go wrong, for you can be sure that unless he is very generous the supplier will, and with complete justification, just shake his head and suck his teeth if you complain of malfunctions having fitted them any other way.

Hostile environment
It's worth reiterating this point about the operating environment because there are few more hostile environments in which electronic components are expected to function reliably than a competition car. Adverse environmental factors include dirt, oil, fuel, water, heat, electro-magnetic interference and vibration. These may be externally or internally generated, or both, such as vibration from on-board components (engine, transmission) or from variously rough track surfaces.

Installation instructions will therefore specify the need to keep components away from obvious sources of liquid contamination, vibration, dust and gravel, direct heat or heat soak (conducted heat), and to place them when ever possible somewhere where they are surrounded by air or where there is an airflow over the component to prevent it getting too hot (typically over about 60°C, although this clearly doesn't apply to temperature sensors that are intended to measure values in excess of this).

Electro-magnetic interference refers to the kind of stray signals that can be picked up from ignition coils, high tension coil and spark plug leads, alternators, starter motors (and apparently telemetry antennae too). Do everything you can to keep any part of your DAS as far away from these components as possible. That statement is made in the knowledge that there is bound to be a degree of proximity between some sensors and potential sources of problems, such as an engine-mounted throttle position sensor necessarily being not too far away from the spark plug leads. But being aware of the potential problems means that diagnosing them if they actually occur will be simpler.

The logger unit
Where you mount this component depends on whether you buy an integrated logger and display unit or a stand-alone unit. In the former case you will mount it where you can best see it, such as on the dash panel or the steering wheel. If it's a separate module you will be able to mount it in a safe, clean position away from other electrical and electronic components, perhaps on a resilient mounting to isolate it from at least some vibration, and, as with any other component, as low down and near the centre of the car as possible for the best mass distribution and a low centre of gravity. Logger units may not be heavy, but every little helps.

As mentioned previously, if the logger unit incorporates integral G-sensors then that's another reason to mount it near the centre of gravity of the car. If G-sensors are fitted near one end or other of the car then they could give false readings. For example, a lateral G-sensor fitted to the rear of the car would give erroneously high values if the car oversteered (that is, the tail swung wide during cornering).

We're now going to take a brief tour around the various sensor types that you're likely to encounter during the course of your relationship with data logging. You may only come across a few of the types mentioned, but it may help to know what else is out there. Sensors come in various types, including speed sensors, position or displacement sensors, accelerometers, temperature sensors,

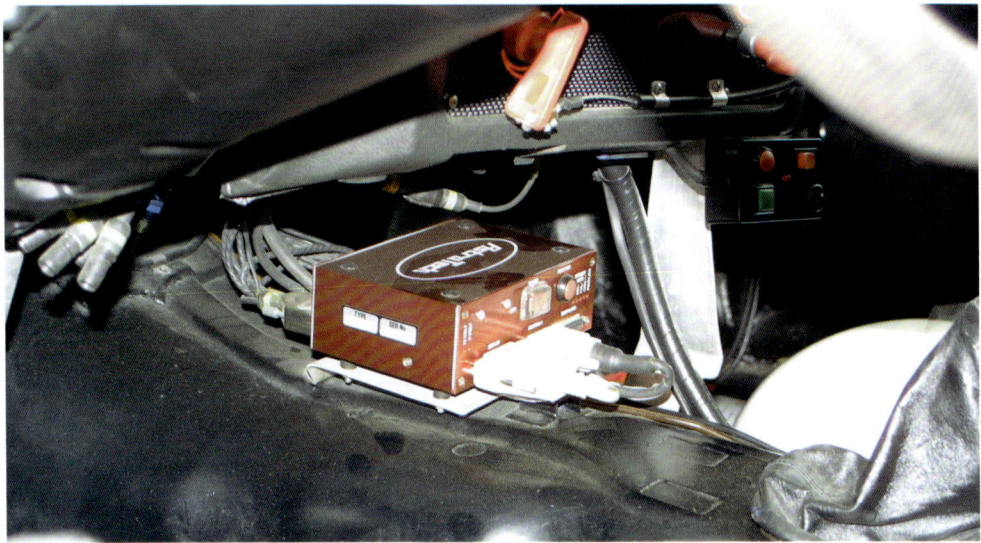

A data logging unit mounted in the cockpit.

pressure sensors, aerodynamic sensors and a variety of more esoteric types that have particular tasks to fulfill. But all sensors have one role in common – they have to convert the parameter that they measure into an electrical signal that the DAS can deal with. Once calibrated for a particular signal the DAS converts this into an output that can be meaningfully displayed. For example, a sensor may convert pressure into volts which the DAS logs, but it then outputs these values to the computer which displays in units of pressure again in order to be understandable.

Speed sensors

There are various types of speed sensor but, for the most part, they have one thing in common – they measure rotational speed. Thus, a road speed sensor actually measures something like wheel or driveshaft rotational speed, and the calibration of the sensor ensures that road speed is what is displayed for analysis. Other parameters which might be measured with rotational speed sensors include crankshaft speed/angle sensors, camshaft speed/position sensors (these two more likely to interface with engine management systems than data loggers), and turbocharger shaft speed. We are more likely to discuss road speed in this book than any of those other possibilities.

In its most basic form, a wheel speed sensor is a magnetic sensor which detects the passage of a magnet (the trigger) fixed to the wheel or the hub, and more sophisticated versions measure changes to electrical properties as one or more ferrous objects (convenient bolt heads, or a ferrous-toothed wheel) pass close to the sensor. Whichever principal is used, the basis is the counting of the measured changes to magnetic or electrical properties which is converted to wheel revolutions and which then, following calibration for overall wheel circumference, is converted to road speed.

Both types are sensitive to the correct gap between the sensor and the trigger, and this needs to be set up carefully at installation. The sensor usually mounts in a bracket attached to the wheel hub or other convenient location, and is held at the correct spacing from the trigger with lock nuts that clamp it to the mounting bracket. As the requisite sensor-to-trigger gap is usually of the order of fractions of a millimetre to maybe one or two milli-

A wheel speed sensor and trigger.

metres, depending on the exact sensor type and specification, brackets and triggers need to be rigid enough to avoid collision-damage that would wreck the sensor.

There is an argument that says if just one wheel speed sensor is to be fitted, then the logical place to fit it is on a driven wheel so that forward traction problems may be diagnosed. However, I favour the fitment to a non-driven wheel – traction problems can easily be spotted by comparing non-driven wheel speed and engine rpm, and I don't think it's worth compromising the information from the speed sensor for the purpose of assessing traction. With either option, there is always the possibility that momentary wheel 'lock-up' under heavy braking can stop the wheel rotating and create an apparently anomalous output, but this in itself can be a useful diagnostic result. It is conventional too to mount the sensor on a wheel which is on the outside of the turns for most of the time on a given course. This avoids logging those moments that can happen when still braking going into a turn ('trail braking') and weight transfer off an inside wheel allows that wheel to lock up. Then again, you might actually want to log this effect, in which case . . . As you can see, there are many tunes to be played, so just think about what you want to achieve and place your sensors accordingly.

Position or displacement sensors

These sensors measure either rotary or linear movement, and typical applications include the measurement of steering angle, throttle position, suspension travel, brake pad and brake disc wear and sequential gearshift selection (that is, which gear is selected).

The rotary types are variable resistance potentiometers which operate like the volume control on your audio system. Changing position through the angular range alters the electrical resistance of the unit. Typical use would be the measurement of brake or throttle pedal movement, or steering angle. Once a sensor in this application is calibrated, for example against the range of steering movement available, the sensor provides a signal that is ultimately displayed as an angle left or right of the dead ahead position. Various angular ranges are available up to a full circle, or 360°. They can be rotated directly, such as on the end of a throttle spindle, or indirectly, for example by a small rubber belt which runs around a steering column and a small pulley on the sensor. O-rings are often used as the rubber belts, and may need some abrasive paper bonding to the steering column to prevent belt slippage. An alternative to the belt-drive is gear-drive. In either case the sensor is mounted on a rigid bracket close to the steering column.

Linear position or displacement sensors

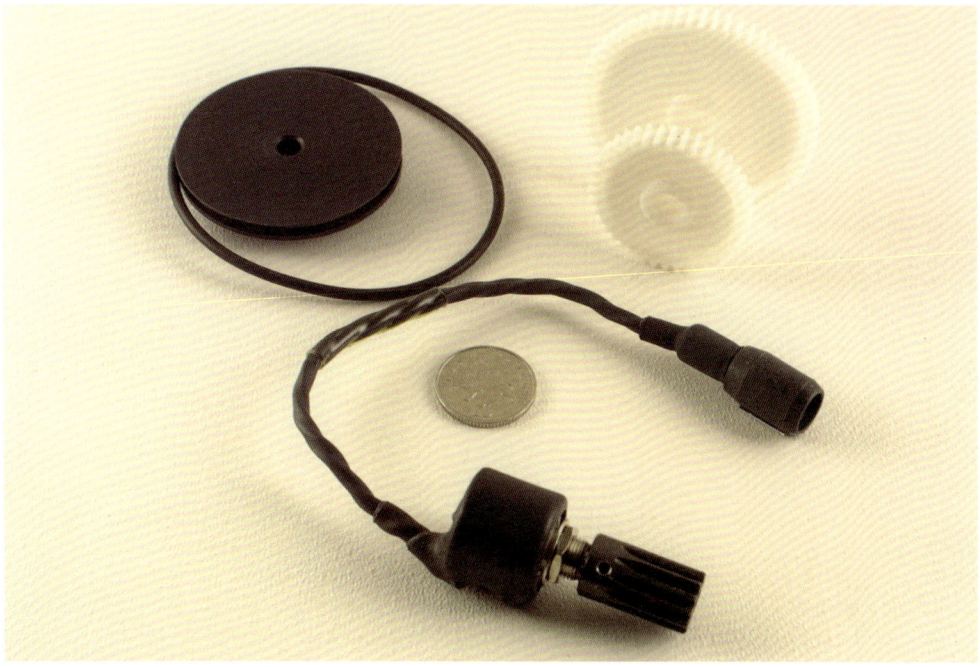

A rotary displacement sensor.

are also variable resistance potentiometers, arranged to provide variable resistance along a range of stroke lengths. Typical applications include the measurement of throttle pedal position and suspension movement. They come in a range of body lengths and offer measurement of strokes from 10mm to 250mm. Typically they have self-aligning rod-end bearings on each end that are used for mounting.

There is another type of linear movement sensor available, called an LVDT, or linear variable differential transformer.

Linear displacement sensors.

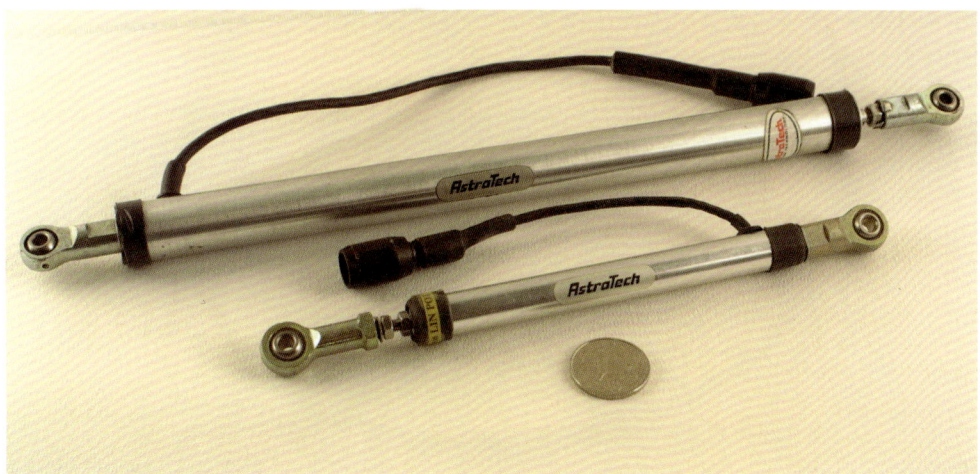

This device works on electro-magnetic principles such that linear movement of a magnetic core induces variable voltages to generate the sensor output. Long strokes as well as very small movements of a few millimetres can be accurately measured, and applications include suspension travel measurement, and brake pad and brake disc wear measurement. Being non-contact devices they have a long service life, but they are more likely to be found in bigger budget race teams. Rotary versions of LVDTs called RVDTs are also available.

Accelerometers

Accelerometers, or G-sensors, are used in competition cars to measure, in effect, the forces created by acceleration, braking and cornering. Acceleration and braking both act along the fore and aft, or longitudinal axis of the car, the former being felt and measured as positive and the latter being negative. Cornering forces are felt as sideways-acting forces, and are thus thought of as lateral. They can of course be generated in left and right-hand turns.

The forces measured by accelerometers are usually reported in units of G. This is the force exerted on any mass by the Earth's gravity, and is a convenient relative unit to use. As persecuted Italian physicist Galileo Galilei discovered when he dropped his massive balls from the Leaning Tower of Pisa in the late 16th century, they accelerated to Earth at the same rate whatever their mass, equivalent to approximately 9.81m per second per second (also stated as metres per second squared, m/s²). This rate of acceleration is said to be 1G, and is a constant on our planet Earth. Thus for convenience, G-sensors, as their name suggests, are calibrated in terms of units of G.

G-sensors contain a small mass suspended on equally small, minutely deflectable beams. Strain gauges (metal foil force-measurement devices) in the beams change their electrical resistance as the mass shifts under the influence of

A G-sensor.

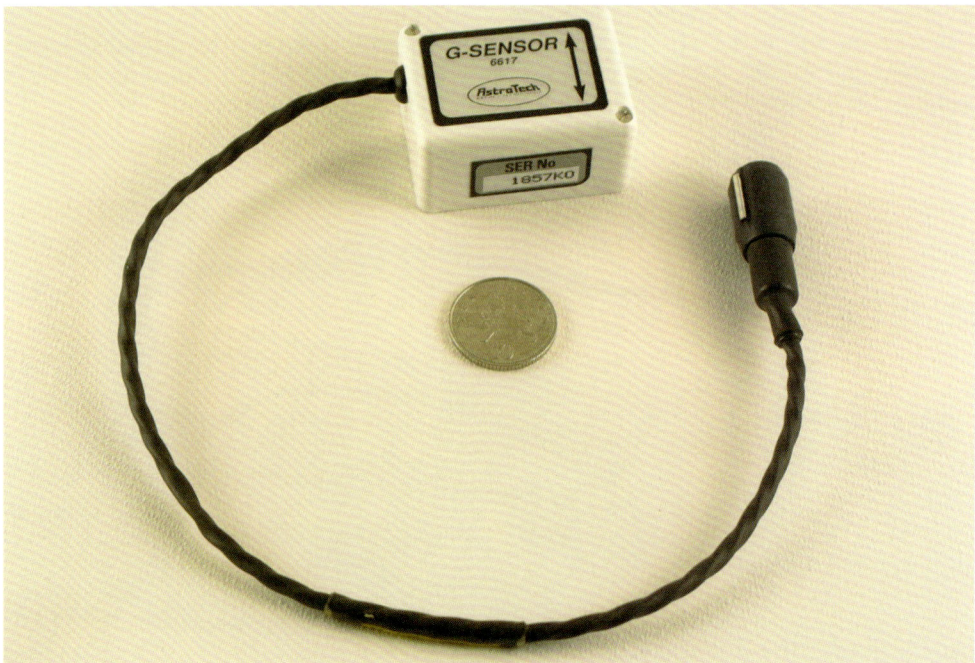

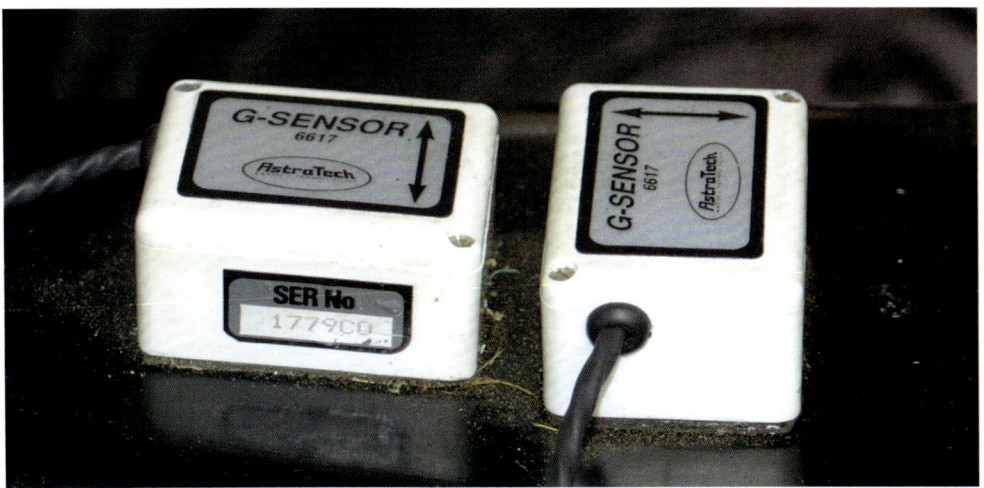

A pair of G-sensors mounted on a firm base in the cockpit.

outside forces that deflect the beams, and the output from the strain gauges is a variable electrical voltage. A simple two-point calibration is done by first resting the sensor on a horizontal surface (checked in two planes with a bubble level) to establish the zero point, and then supporting it so that the measurement axis is pointing vertically downwards towards the ground, which will establish the 1.0G point.

It will be obvious from this that installation of a G-sensor must be done with the car on a completely level surface, and the sensor itself must be installed with the relevant axis level in the car. As mentioned previously, it is important to mount the sensor near the car's centre of gravity (CG). If it is mounted ahead of or behind the CG then it will pick up errors due to yaw accelerations (for example tail-out or oversteer moments). Similarly, if it is mounted above or below the CG then it will pick up errors due to roll or pitch accelerations (strictly speaking, a car yaws, rolls and pitches around geometric centres that rarely coincide with the CG, but the principle here is valid).

Accelerations due to vibration are another source of potential error and 'noise', and anti-vibration mounts are usually recommended, although mounting with industrial Velcro is said to be satisfactory. Clearly, a mounting medium which is too soft or flexible will incur its own errors. Avoid mounting a G-sensor in the middle of an unsupported panel which is bound to vibrate.

G-sensors of different ranges and levels of precision may be obtained; your specific requirements and your budget will dictate which, if either, type you obtain. For most general purpose applications a range of +/- 5G suffices. It is also possible to buy single-axis as well as two- and three-axis G-sensors.

Temperature sensors

There are various types of temperature sensor available for the measurement of fluid (water, oil, fuel), air (ambient, engine airbox, inlet manifold), and hot gas (engine exhaust) temperatures, as well as contact and non-contact devices for measuring tyre and brake disc temperatures. Temperatures up to 1,200°C may be measured, although the applicable sensor range would be selected according to the parameter of interest. Most of these different sensors are based on a thermocouple or similar component whose electrical properties change predictably with temperature.

The circumstances that dictate whether

it will be useful to log any or all of these parameters will depend on the type of competition you are involved with, and maybe on the development phase you are undergoing. Unless you have plenty of spare logging channels and a DAS with lots of available memory you are unlikely to want to log oil and water temperature unless you are checking or refining the efficiency of the car's cooling system, or you are perhaps running in a long distance event. Similar reasoning may be applicable to some of the other temperature parameters that you could log – there may be specialised circumstances that require these functions.

If you have a DAS with, say, four to six channels you are probably unlikely to log any of these temperatures on a regular basis, but that's not to say that you could not do so should the need arise. With the purchase of a suitable sensor you could disconnect one of your regular sensors and connect up, for example, a water or exhaust gas temperature sensor to that channel while you researched what's going on during testing or an event.

Tyre temperature measurement by non-contact infrared radiation sensing is interesting because it has two facets. With an array of sensors across the width of a tyre it becomes possible to analyse the behaviour of the tyre and suspension by examining tyre temperature profiles during running, as opposed to after coming to a halt in the pits. But of greater value is the reliability and safety issue that arises when a tyre starts to deflate and overheat, which at best is likely to cause an unnecessary visit to the pits, possibly a non-finish, and at worst a potentially dangerous accident. Of course, this is only of any use when linked to a telemetry system that relays the data to the pits in real time. Thus its use is limited to the upper echelons of motorsport.

Pressure sensors

Pressure sensors are used to measure such parameters as engine inlet airbox pressure, turbocharger or supercharger boost pressure, fuel pressure, oil pressure and hydraulic pressure in actuators or hydraulic fluid circuits like the brakes. Clearly, these different applications require different pressure ranges from the low end in the case of engine airbox pressures (to perhaps 25psi or 172KPa), to the upper end in hydraulic lines (to perhaps 3,000psi or 20.6MPa).

Various principles permit the measurement of pressure, and sensors include the diaphragm pressure type. These have metal foil strain gauges bonded to a diaphragm. When there is pressure on the diaphragm the strains in the foil gauges change, and a change in electrical output which is proportional to the pressure takes place. Pressure can also be measured by piezo-resistive sensors. Piezo-electricity (from the Greek, literally meaning 'press' electricity) is generated by certain crystal and ceramic materials when they are subjected to pressure. These materials are suitable media for inclusion in compact pressure sensors, and can be made to measure a wide range of pressures from 2bar (29psi or about 200KPa) up to 250 bar (3,625psi or about 25MPa).

Aerodynamic sensors

These may either be low range pressure sensors or air speed sensors. The changes in pressure around a competition car (even one that creates substantial amounts of downforce) are small relative to the pressures in oil, fuel or hydraulic lines and so pressure sensors that are of use in aerodynamic data acquisition must be able to measure in the range of zero to just 1 or 2psi (70 to 140mbar or 7 to 14KPa). The piezo-resistive principle can be used for these low pressure measurements.

For air speed measurements, two methods can be used. The first uses the so-called pitot tube device, which is a small bore tube that protrudes above the surface of a competition car body and faces into the airflow. By evaluating dynamic pressure and using Bernouilli's Equation (which relates local pressure to air velocity) it becomes possible to measure air speed. The other type of device is a small, lightweight rotating vane mounted on low-friction bearings. This has a similar function to the speed indicators we looked at earlier in that, following appropriate calibration, it converts rotational speed into linear speed.

Other sensors

There are various other types of sensor available, such as strain gauges and load cells to measure suspension forces, laser ride-height sensors, shaft torque sensors and so on. But as these really are for the seriously well-funded top professional teams, hopefully you'll forgive me if I make the assumption that you probably won't be using them. We will take a brief look at them and how their data is used in Chapter 13, but in truth there is enough information to get to grips with from just the basic sensors, and that's what we're going to concentrate on throughout this book.

Wiring and connections

If this section seems like a repeat of earlier information, well, it is. But we've all suffered from components or devices which have not worked correctly at first, and you have to admit, it's usually down to installation. Hence the need to repeat some of the basic advice that DAS suppliers offer. In fact it has been said that most problems with any electronic systems on competition cars or modified road cars are caused by faulty installation, in particular with the physical installation of wires and connections. So make all your connections secure ones by following the supplier's instructions. If they sell a special crimping tool for the job then buy one. Don't penny-pinch and use those dentists' pliers you picked up at the local car boot sale! When visiting one of the leading DAS suppliers once, I confess I was surprised to learn that connections were made by crimping, not soldering.

Crimped connections are, apparently, better electrically, and soldered joints can also be brittle enough to fracture when exposed to constant vibration. This same company actually went to the trouble of calibrating all its crimping tools in order to produce the same, top-quality joints every time. Makes you think, doesn't it?

The location of sensors is another potential source of problems, and some of the environmental problems that exist on competition cars were highlighted earlier. Sensors need to be secure and rigid enough so that vibration and G-forces do not bring them into contact with moving, rotating or hot parts that they are not supposed to be.

Cables must be routed and secured with care. Again, the obvious things to avoid are moving or hot parts, but also route them away from areas where they might get trapped or crushed, say during an engine change. Keep them away from all fluids or areas where fluids may spill, leak or accumulate in the event of problems with a fluid-carrying system. Also keep them away from other electrical cables, especially high-current or high-voltage systems like the ignition circuit or high-tension plug leads. Follow the supplier's instructions and advice closely in this regard.

Don't run cables too taut in case there's tension on the connectors which might, with the help of vibration (a commodity in plentiful supply on a competition car) disconnect your DAS for you in mid-run.

One of the most important connectors in your DAS system is the download socket. This obviously needs to be readily accessible by hand and with the plug provided. It must also be physically strong and rigidly mounted so that repeated connection and disconnection doesn't cause any problems such as the socket mounting fracturing or bending in a way that prevents access with the download plug.

Chapter 3

Basic analysis tools

LOVE THEM OR loathe them, computers are an everyday, common-or-garden fact of life that most people have to experience at some time or other, to a greater or lesser extent. I was something of a late-starter in computing, but very quickly became hooked on the benefits, and now couldn't do without them. Once I'd got my first computer it wasn't long before it was connected to the racing car I was using at the time in order to extract data from it.

Taking the plunge into computing!
If you're going to benefit from data logging you will need to have at least a basic ability in the use of a personal computer (desktop or laptop variety – we'll discuss what's best later). If you are a complete computer novice, have no fear, Appendix A will offer some guidance to the absolute basics. So, if you've always thought the cable on a computer's throttle pedal was too short to reach the floor, go straight to Appendix A without further ado. If however you were not tricked by the preceding sentence and you not only know what a mouse is but also what it does, you're experienced enough to get straight into data logging!

Actually that's a minor exaggeration – my four-year-old nephew knows what the mouse on a computer does already, but I'm guessing he's only just beginning to realise what understeer is going by his pedal-kart technique, and therefore probably wouldn't benefit much from logging steering angle! The experience of a friend's father however serves to remind all computer-literates of the difficulty in grasping some concepts of computing. He simply couldn't comprehend how a word that had just been typed 'on-screen' could then just as easily be deleted. Painting correction fluid on to a paper page was one thing, but removing letters and words from a screen just by pressing a button was quite another. So I'm going to try not to make too many assumptions about any foreknowledge of computing in this book, and that's why there's a section in Appendix A that might get you over some of the initial hurdles if you're a novice in the subject. There are of course plenty of specialist books on computing as well as basic courses you can take, but I thought it might save you a bit of time if this book included some of the basics. This inevitably means some readers will perceive I am treating them as simpletons, but as I am told there are lots of people out there who want to use data logging, but don't know how to use computers, bare with us, and skip the bits you already know enough about.

What I am therefore going to do in this chapter is to assume you know how to start a computer, how to load and start programs and how to use their features and functions using keyboard and mouse commands. These days most personal computers (PCs) use software based on Microsoft Windows, and whatever you may think of Microsoft's virtual monopoly of the computer 'operating system' market, it is actually hugely advantageous that most of us use the same type of software. Not everyone has the same type of computer, and if you're in the market for a computer to enable you to take up data logging but you don't know what to buy, it may help to consider the pros and cons of the two basic options.

Portable or not?
The two options are desktop or laptop. The desktop computer is not portable, the laptop is. Other than that their functions are pretty similar. The first thought that almost inevitably comes to mind is that a portable computer is going to be more use if all you need to connect it to your competition car is a short cable. For this simple reason, laptops are the favoured choice for data logging. However, excellent though they are, their display screens are not of the same quality, clarity or visibility as those on desktop PCs. Having typed that sentence, technological advances have probably made it an obsolete and incorrect statement already, but as things stand I believe that desktop PC screens are easier to work with. Clearly desktop PCs are less portable, but you do see some teams who set them up in the pits at race meetings, having probably transported them in padded cases.

Desktop PCs have other advantages too. For example they usually have a nicely spread out keyboard compared with that of a laptop, which is necessarily a compacted version, and a desktop PC is more likely to have all the disk drives you need contained within the base unit or tower unit, although laptops are remarkably well packaged in this respect now, especially at the upper end of the price range.

Laptops have another great advantage which is intrinsically tied in with their physical portability and that is because they run off internal batteries they do not require mains electric power. So other than the download cable that connects the DAS to the computer, you can use a laptop without any other cables. This remark holds true until such time as you start wanting to connect additional external devices like supplementary disk

Laptop or desktop PC? Both have their advantages.

drives, for example to make additional copies of your data (always a smart move which we'll discuss in more detail shortly). Of course, a laptop's batteries will need recharging after a few hours use (my current laptop, purchased in 2001, typically offers $2\frac{1}{2}$ hours of use from a fully charged battery), which will require mains power. Another smart move is to buy a spare battery, but be warned, they cost a lot more than ordinary household batteries.

You could also consider buying a laptop for trackside use, and a large, clear monitor to connect to back in the workshop/office/home. One professional race engineer advised a 21-inch or 22-inch monitor to connect to the laptop at the circuit – 'the bigger the better' he reckoned because it makes things so much clearer. That would give you the best of both worlds, but presupposes a dry pit garage or support vehicle in which to site such a big monitor.

What specification?

It would be a complete waste of time to spell out exactly what specification your computer should have because they are rapidly advancing all the time. The only sensible thing to say is that the DAS you choose will govern the amount of computing power you will need. The supplier will be able to tell you what type of processor is needed, what type of display, what operating system is required (usually though not necessarily translated as 'which version of Windows'), how much working memory (RAM, currently still measured in MB or megabytes as this is typed) is needed, how much free storage memory is needed for the DAS software – also measured in MB or GB (gigabytes) of hard disk space – and so on.

In general with computers I think it pays to spend as much as you can on a decent system, provided that is you don't go back a week later to check out the price on the model you've just bought, because it will be cheaper than when you bought it! That seems to be an immutable law. The desktop computer I bought most recently cost me £1,500 (about $2,100 at the time of writing) yet three years later I could buy a PC with the same specification for less than half that price. By stretching my budget as much as I could to buy what was then a fairly high spec, my PC is still good enough to run current software without needing a major upgrade (or replacement) and hopefully will see out another year or so before that is needed. If you buy the cheapest available, assuming you are actually still able to run your DAS software, it may well run irritatingly slowly. Again, ask your DAS supplier for advice – maybe a cheap second-hand laptop will do the job, in which case you can save some money, but you need to at least match the equipment to the current software, and maybe build in some 'future-proofing' if you can afford to.

Hard copies?

A printer is a useful addition to a computer for doing most things, including data logging. Sometimes it's just easier to see details on hard copy printouts than on a screen, however good your screen might be. You may not take the printer with you to events, but if you want to then you have the equivalent of the 'desktop versus portable' choice again. But of more use than portability is colour printing, so that all those different coloured wiggly lines that represent your data can be printed in their individual colours and distinguished from one another. Fortunately, colour printers can be bought for amazingly small sums of money these days so hopefully the cost will not be a problem. Many home desktop computer systems come 'bundled' with budget colour printers nowadays, and it's an easy job to disconnect it from the desktop and connect it to a laptop if that's the route you go.

Loading software

It should be readily apparent from the manufacturer's instructions how you should go about loading the DAS 'software', that is, the program that enables you to extract data from the on-car DAS and analyse it, on to your computer (henceforth referred to as PC because it takes less typing, but take it to mean desktop or laptop). Going by peoples' varied experiences this isn't always the case, but hopefully you'll be able to rely on the instructions supplied. The software will either come on a 3.5in floppy disk (they're not floppy, but don't let that be a concern) or, more likely nowadays, a CD. (In years to come somebody will blow the dust off an old copy of this book and ask 'what's a CD?' but I'm guessing they'll be around for a while yet . . .).

You will then need to insert whatever type of disk on which your software came into the appropriate slot or drive on the computer, and either wait for the thing to magically start loading by itself, or else type in the stated command, such as 'setup' or 'install' in the requisite space and then tap the return key, or 'click' on an 'OK button' (go to Appendix A if this bizarre phrase is meaningless to you!). This should initiate the loading process, in which the program files are copied on to your PC's hard disk, and once the process is complete it allows you to run the DAS software from the PC in future without having to reload each time.

If your chosen DAS software runs under Windows and is any good at all, there will be an icon, (a little coloured symbol representing the DAS software) on your PC 'desktop' (the opening screen when your computer has started up). So, to start up the software you just double-click on this icon which will present you with the program's opening screen.

If you do not have an appropriate icon visible on the screen and you're sure you've loaded the software correctly, then shame on your DAS supplier for failing to make things easy for you! Not to worry though. What you'll have to do is to click on the 'Start' button on the bottom left of your screen, place the pointer over 'Programs' in the list that pops up on your screen, then search through the listings until you find the name of your DAS software. Double-click on this and the software should start up to the opening screen. If your DAS software does not run under Windows, follow the supplier's instructions to access it via a somewhat different, but essentially still very simple route (see Appendix A).

Download and saving data

So the DAS software is up on-screen, staring you in the face but somehow not psychically detecting what you want it to do and performing that duty for you! You have to provide some more instructions, with mouse or keyboard selections, and these will obviously depend on your chosen system. The next step, assuming now that you have physically installed your DAS and sensors on the car and you've driven it around to log some data, is to retrieve, or download that data to the PC.

Retrieving data from the DAS and loading on to the PC is a simple procedure, whatever DAS you have. It just involves connecting the download cable to the relevant socket (known as a port in computer-speak) on the back of the computer and then to the download socket on the car. You then need to follow the instructions for your particular system, which might require some button pushing, or maybe making some selections with the mouse or the keyboard on the PC to activate the download sequence.

The DAS software may prompt you via on-screen messages to enter some memo details in boxes provided for the purpose, such as the name of the venue you are competing or testing at, the date, and maybe a few other pertinent details like the car and driver names, weather and track conditions, car set-up details and so forth. All of this will help you when it comes to later analysis and you need to remind yourself of that particular run, race or session. Then you will probably have to select a button that says something fairly helpful like 'Download now', and the data will hopefully start streaming up the cable from the DAS to the PC. This process may take a few moments depending on how much data your system can log, and on the rate of data transfer the system permits.

When all the data has been transferred, you will probably then be sitting in front of a screen full of coloured wiggly lines, and this is where the fun and the accumulating of vast amounts of new knowledge begin. But first there's some administration to do. That probably sounds very boring, and in truth it is, but if you don't do it you may very well kick yourself later, or possibly the cat which would be very unfair if it's you that failed to securely save this precious information. The data that you have just transferred to your PC is held initially in the computer's RAM (random access memory). Why do you need to know that? Because whatever is held in RAM disappears when the computer is shutdown and turned off, so if you shut down your PC now, or its battery goes flat, or it just suffers one of those infuriating 'crashes' from which they do suffer from time to time, then your data will be lost.

To avoid this embarrassing and annoying situation, what you have to do, and I really do recommend you do it at the earliest opportunity, is to save the information in a file. This is the PC's electronic equivalent of putting a valuable piece of paper safely in a real file so that it doesn't get lost and you can find it easily later. Usefully, the software allows you to give names to files, analogous to those tabs you write on in a filing cabinet's hanging file system, for example 'Monaco GP Session 1-0501', giving a quick reminder of place, date and time. The file is then a permanent record of that particular set of data. Permanent is a dangerous word to use with computers though. One day, when I tried to start my PC, its power supply unit and hard disk both failed simultaneously, for reasons I never did get to the bottom of. At that point, and over the next few weeks, I discovered how much of my valuable (to me) information had gone up in smoke with the hard disk. And guess what? I hadn't 'backed up' some of the crucial stuff.

Backing-up data is another habit that I strongly recommend you get into in order to avoid the inconvenient mess I had following my hard disk's demise. It means making copies of your files on floppy disks, or CDs or whatever medium your computer allows, also at the earliest opportunity. Of course, I had read and listened to this same kind of advice before my PC failed, but I had only done a partial job of following that advice. Since my PC's failure, I am now fairly zealous about making copies of important files. So take it from one who suffered the consequences of failing to follow good advice, and back up your files! See Appendix A for more help on this if you need it.

You may now safely begin the process of looking at your data, secure in the knowledge that you have copies of it all over the place – well, two copies anyway. But before we get into the generalities and details of data interpretation, it's also worth remembering that you can make hard copy printouts of the data in several different forms, depending on your chosen data logging system. You can obviously print the graphs, which are perhaps of most interest. But analysis software often allows different types of graphs to be shown too, such as bar graphs (or histograms if you prefer) and other graphical representations showing particular types of information. You can print out the raw data on which the graphs are based if you like poring over vast tables of numbers. And you can print out the basic memo reports showing where the data relates to and so forth. Some of this will provide useful, rapidly accessed information for when you are away from your computer or you just can't be bothered to start it up and it's quicker and easier just to reach for your paper record files!

Basic graph interpretation

If a picture is worth a thousand words, then a graph is probably even more efficient at portraying certain types of information. I've always been a fan of graphs, and have perhaps been guilty of using them to excess sometimes in presentations and articles. It's easy to forget that it takes more than just a moment for someone looking for the first time at a graph I have prepared, and am therefore totally familiar with, to fathom what it is demonstrating. Graphs are not everyday forms of information presentation, unlike words and pictures, and it would therefore be wrong of me to assume that readers of this book are necessarily familiar with what graphs are able to tell us. Once again therefore, would the graph-conversant amongst you bear with us while a brief study of the basics is made.

The line graph

Graphs are a very convenient way of visualising a mass of numbers, or data. They are generally much more meaningful than tables of data and make it easy to see what has happened to a parameter of interest. Take the most basic form of graph, the line graph. This format could, for example, be used to show how the value of shares in Interplanetary Tours.com has been nosediving since the annual report admitted that the scheduled inaugural tour date was slightly optimistic. If such a thing existed, that is. The format of the graph in such a case allows a parameter, the share value, to be plotted against time, so you can see at a glance how things have changed over a period of time. See Figure 3-1.

In graph terminology, such a line graph is constructed with a pair of axes (pronounced 'axeese', it's the plural form of axis, lest you thought tree-felling implements were involved). The horizontal axis along the base is known as the x-axis, and the vertical axis up the side is the y-axis. In the case of our mythical share value fluctuations, the share value, in pounds, dollars or whatever, is shown on the y-axis, whilst time is shown on the

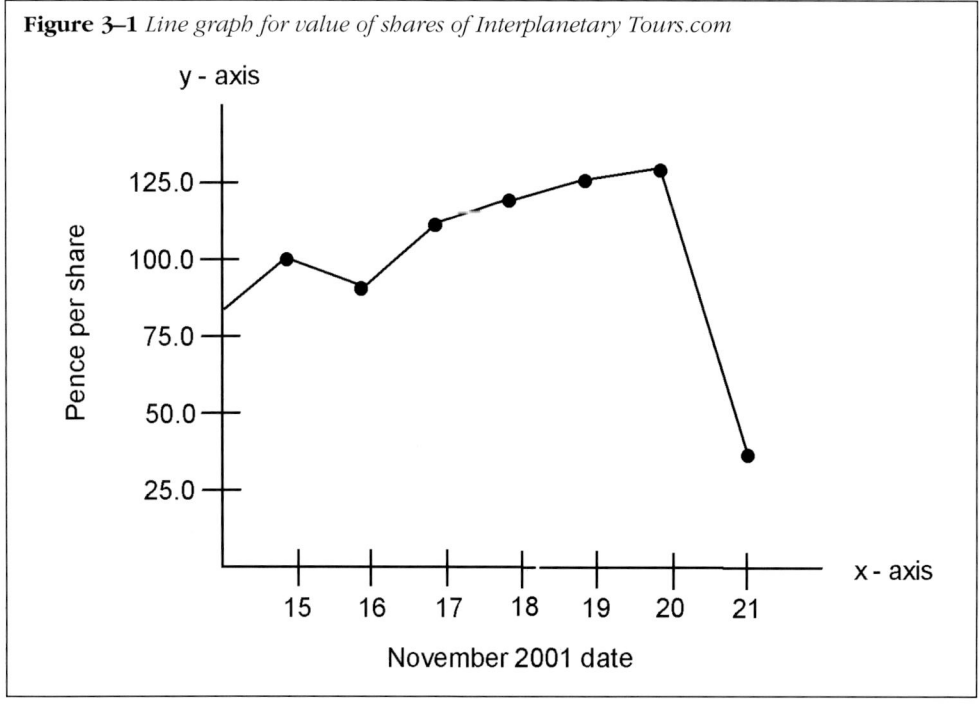

Figure 3–1 *Line graph for value of shares of Interplanetary Tours.com*

x-axis. In this case, time might be shown in days or weeks – it is shown in days in the example here.

Thus the wiggly line in the body of the graph is actually constructed by entering each day's share value as a discrete point, and then joining up all the points to form a line. Straight away it is possible to see how the value changed from day to day, and by annotating the graph with key events, the reasons for some of the changes can be made clear. The other thing that becomes obvious is how longer term trends are evident as the brain seeks out, and frequently finds, underlying patterns in the data, such as the trend towards a higher value in the early part of our mythical example graph.

Getting back to the slightly less surreal world of competition cars, a similar type of graph to the share value-against-time plot would be an engine rpm-against-time plot (see Figure 3-2). The graph is constructed in just the same way, except rpm takes the place of share value on the y-axis, and seconds replace days on the x-axis. The wiggly line in the graph body is now constructed by the rpm values sampled 10 or 20 times per second (or whatever rate your DAS samples at), and the line is constructed by your DAS software by joining the dots, so to speak, and presents you with something you can interpret easily.

The software is also clever in that it looks at all the data, checks what the maximum and minimum values of rpm are during the session you have been logging, and adjusts the y-axis accordingly. Thus, if your engine never exceeded 6,850rpm it would be rather pointless setting the upper limit of the y-axis to 30,000rpm because all the detail you want to see would be compressed. The software sets the upper limit to a convenient 7,000rpm and you get all the detail you need in clear view. You can usually 'scale' your data manually too, to fit it on the screen the way you want.

That's just one example of how a line graph might aid data analysis. Another form of line graph replaces time on the x-axis with distance. If you think about it, this is of much more value because if

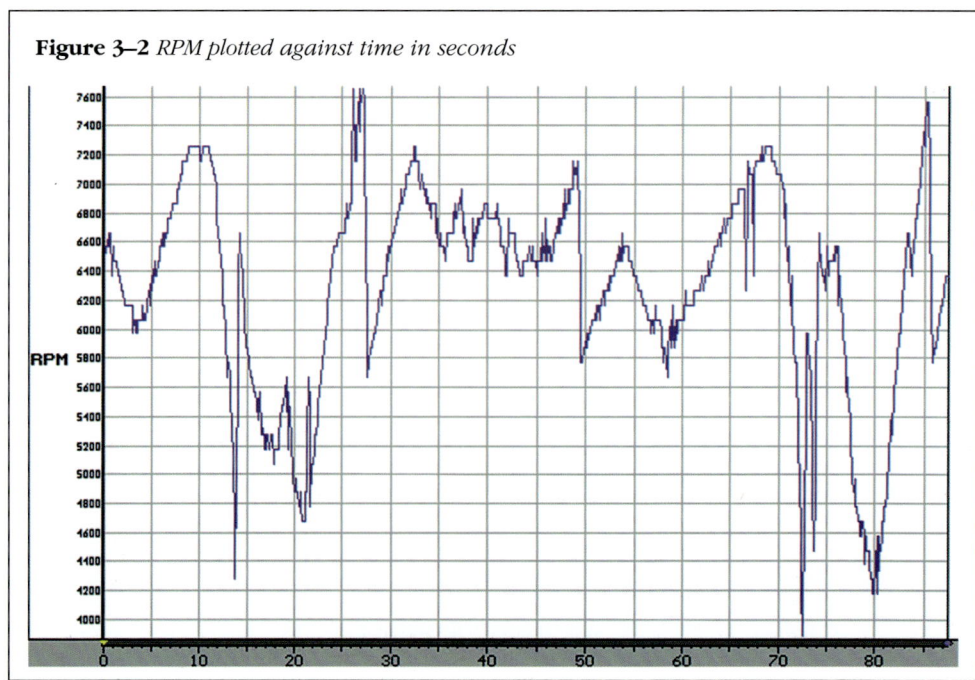

Figure 3–2 *RPM plotted against time in seconds*

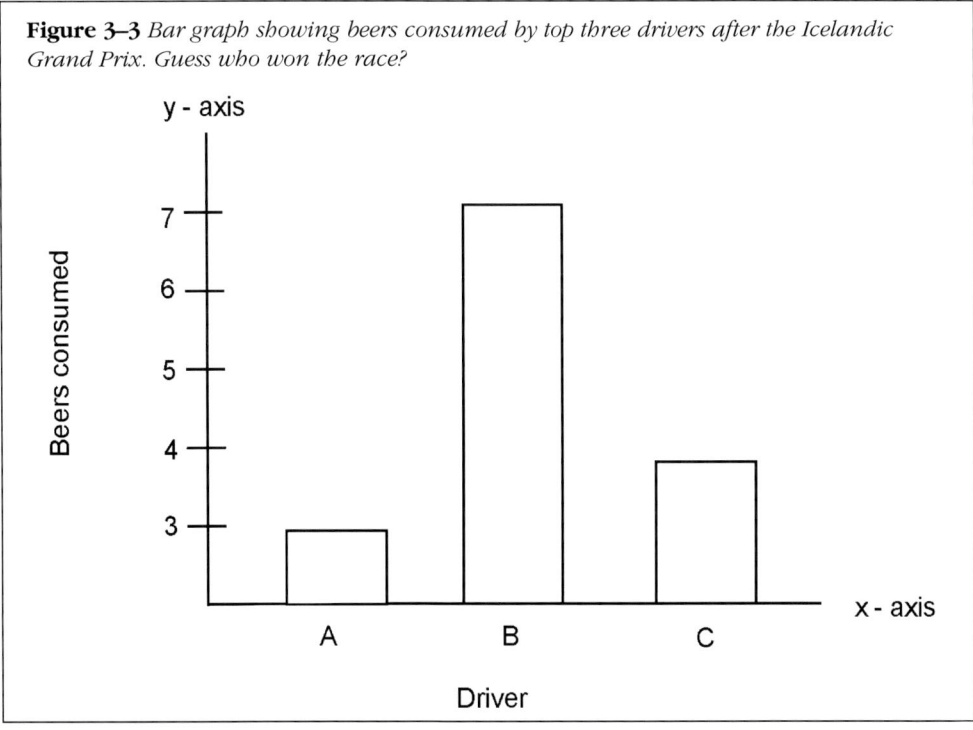

Figure 3–3 *Bar graph showing beers consumed by top three drivers after the Icelandic Grand Prix. Guess who won the race?*

you're driving around a given circuit the one thing that doesn't alter is the lap distance (although you could make it longer with off-course excursions I suppose). What you are usually going to be analysing is where things happen in that fixed distance. So DAS software also allows you to view your data with distance plotted on the x-axis. This is a case of horses for courses – there will be circumstances where you are more interested in viewing some parameters against time, or when comparisons of different data sets become more useful if measured against time rather than distance, and vice versa.

The bar graph

A bar graph is a means of illustrating the frequency with which a number of events have occurred. For example, it might well be possible to demonstrate how many beers the top three drivers consumed after a race, and possibly try to interpret that information in relation to their positions on the podium. Another name for a bar graph is a histogram, so maybe we'd better stick to that title – all this talk of bars could be distracting.

Of course, the bar graph is so-called because such a graph consists of a set of bars, arranged either vertically or horizontally. Each bar's length demonstrates the frequency of the parameter it illustrates. Going back to the beer consumption example again, the y-axis could show the number of beers consumed, whilst the x-axis would show drivers A, B and C if anonymity needed preserving, or any other label that was appropriate. See Figure 3-3.

A more serious application might be to show how long the throttle was held at certain openings. For example, when lapping a circuit full throttle would be employed some of the time during the lap, whilst at other other times, part-throttle openings would be employed. By using the software to split the data into throttle-opening 'brackets' such as 90 to

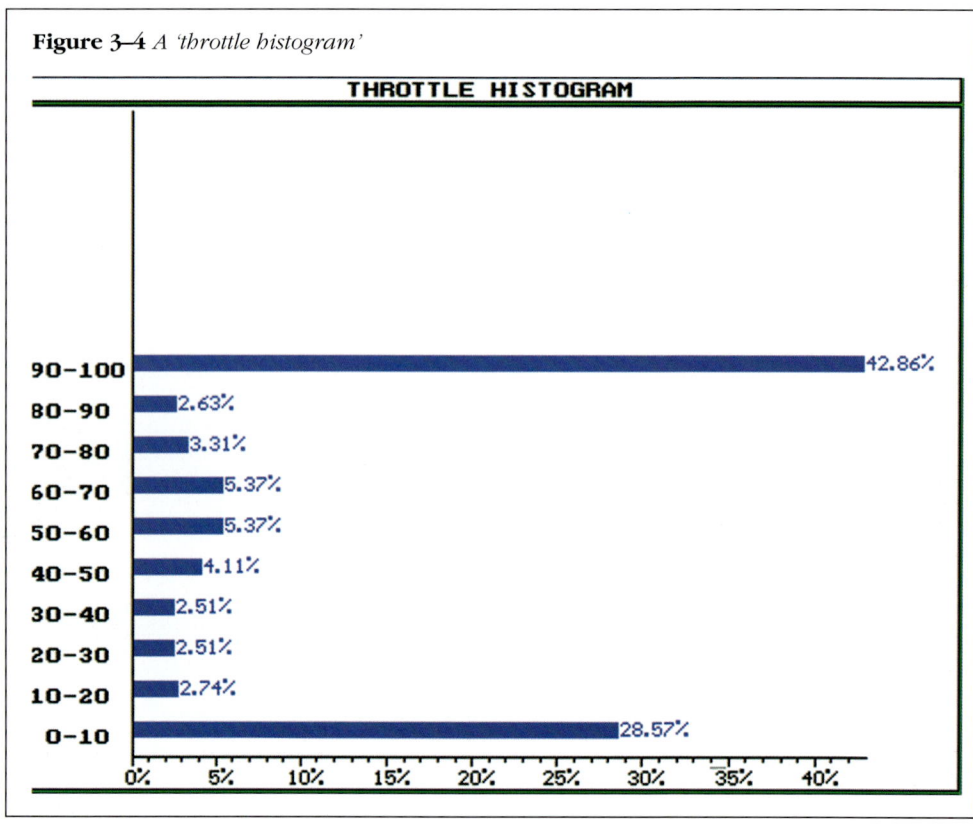

Figure 3-4 A 'throttle histogram'

100 per cent throttle open and so forth, it can then calculate the amount or the proportion of time spent in each of those brackets and display that information as a histogram. See Figure 3-4. Note the phrase, 'amount or proportion of time' in that last sentence. It might be possible to show the actual time in seconds spent in each bracket, or the proportion of the total time. The graphs would look much the same in each case, so that would just be a matter of preference. The same could be applied to the beer consumption graph – you could either plot the number of beers each driver drank, or the percentage of the total number of beers that each driver had consumed. The graph would be the same shape in each case, only the y-axis labels would differ. Again we'll go into the whys, wherefores and uses of this type of data analysis later on.

X-Y graphs

X-Y graphs are used to simultaneously plot two related parameters on one graph. You might perhaps try to plot Interplanetary Tours.com share prices versus the number of beers consumed by known shareholders. You'd get a fairly interesting plot if the shareholders drank as heavily in celebration of high share values as they did in the drowning of sorrows over collapsed values.

However, in the competition car context, a favourite use of the x-y graph is in plotting longitudinal and lateral G on the same graph. Another use might be to plot speed versus rpm to produce a sort of gear chart (see Figure 3-5) – other interesting points may emerge from such a plot too, and we'll go into that later on in the data analysis. Suffice for now to be aware of the type of plot and what it represents.

Figure 3-5 *An X-Y graph of speed against rpm*

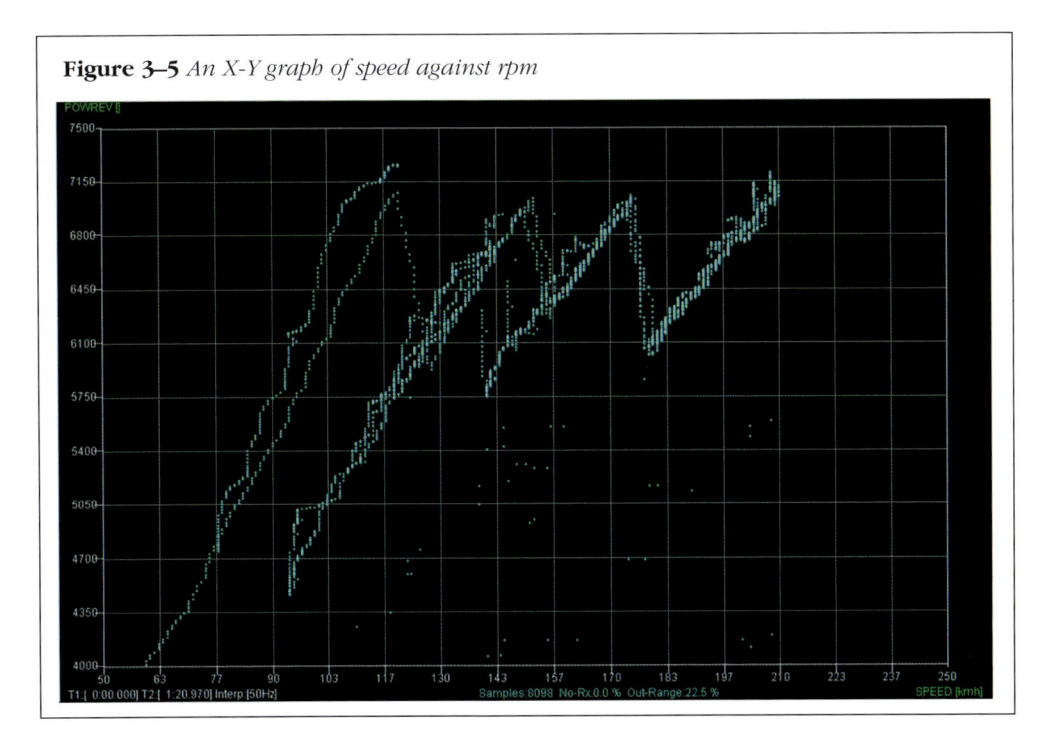

The three graph types explained here are probably the most common ones you'll come across during data analysis, and they are put to many and varied uses. There are other types of chart or graphic that crop up too, such as lap segment charts showing the time differences on the same segments of different laps, and track maps which show the position of the car around a circuit and possibly much more too. We'll deal with these as we get to them.

Chapter 4

Rpm analysis

THE FIRST FORM of data logging that became widely available was engine rpm recording, using what the manufacturers call 'intelligent tachometers'. Unlike their less cerebrally-developed mechanical predecessors, which had telltale needles that simply lodged at whatever the maximum rpm had been during the running following the last reset (which you could only hope wasn't just as the driver came back into the pits . . .), these devices contain an electronic memory that records the engine rpm. The data can be accessed in a variety of ways, including linking the rev counter memory to a PC to download the information for detailed analysis. These systems, now refined somewhat, are still more than capable of yielding a surprisingly large amount of information, not just about what the engine has been doing, but with a bit of focused study and thought, also about the car and the driver . . .

The recording tacho essentially logs the last few minutes of engine running time at a rate of 10 to 30 times per second, depending on how old the tacho is and whom you bought it from. The amount of memory provided governs the length of running time that can be stored, and the memory is basically like a continuous loop tape recorder, recording the latest data over the oldest as the memory becomes full. So, for example, a basic starter system might offer enough memory for around seven or eight minutes of engine running time, and with upgrades or by starting further up the available product range, you can have around 25 minutes of memory. This is probably enough for most club races, whilst the smaller memory capacities would be ample for hillclimbs, sprints, drag racing and other shorter events.

When using one of the earlier recording tacho systems on a hillclimb car some years ago, we had a separate power switch rigged up to the tacho so that it could be turned on just prior to a run and off again just after the finish in order to compact as many runs as possible into the 6.7-minute memory. This prevents irrelevant data from being recorded – you don't really want to know about the rpm whilst warming the engine up or driving out of the pitlane – and makes it easier and quicker to select the data you're actually interested in analysing. Some current systems offer the ability to initiate logging with the press of a button to fulfil the same end.

The basic components of an rpm logging system are a rev counter with internal or external memory, some buttons to reset the memory, to put in markers and as mentioned, to initiate logging, and optionally a download socket and cable to link to your PC. The download option is just that, an option, because not everyone has a PC and it is possible to use these clever rev counters simply as playback devices. This facility is used after returning to the paddock, and with the pushing of appropriate buttons, the rev counter replays, via its own needle, the data stored in its memory. It can be played back in real time that is, at the same rate at which it was logged, to give you an exact replay of what you'd have seen on the tacho if you'd watched that instead of looking where you were driving. Or it can be played back speeded up or slowed down if you so choose.

The replay facility is a useful way of extending the ability of a mere rev counter, but data cannot be saved for later analysis and it soon gets recorded over, at which point it is lost. Surely then it's far better to be able to download this potentially extremely useful information into a computer, and to then be able to save it and analyse it immediately or later if you so desire. Yes there's the extra cost of a computer, and also the software to facilitate downloading and analysis, but trust me, if you're interested enough in data logging to have bought this book then you won't regret the extra outlay on a PC and the relevant analysis software. (Your bank manager might regret it, but who's in charge here?!)

Data interpretation basics

Following on from the fundamentals of graph interpretation explained in the previous chapter, it's now time to take a look at our first competition car data logging traces to see what is being represented, and to begin relating that to what is happening with the car and the driver. Once again I'm aware that not all readers will know exactly what the wiggly lines we're going to study here actually represent, so the graph and data conversant must grin and bear it whilst we start from the basics.

The output you usually get from an intelligent rev counter is a record of engine rpm against time. In graph form, this is represented with time along the horizontal or x-axis, and rpm on the vertical or y-axis. A typical rev counter graph plot is shown in Figure 4-1. This one is actually from a run on a hillclimb, but it could just as easily be a section of a race circuit, or a number of other competition environments. The chart shows thousands of rpm, or krpm as it is shown, plotted versus time in seconds. The rev counter was set to 'record' by the driver pushing

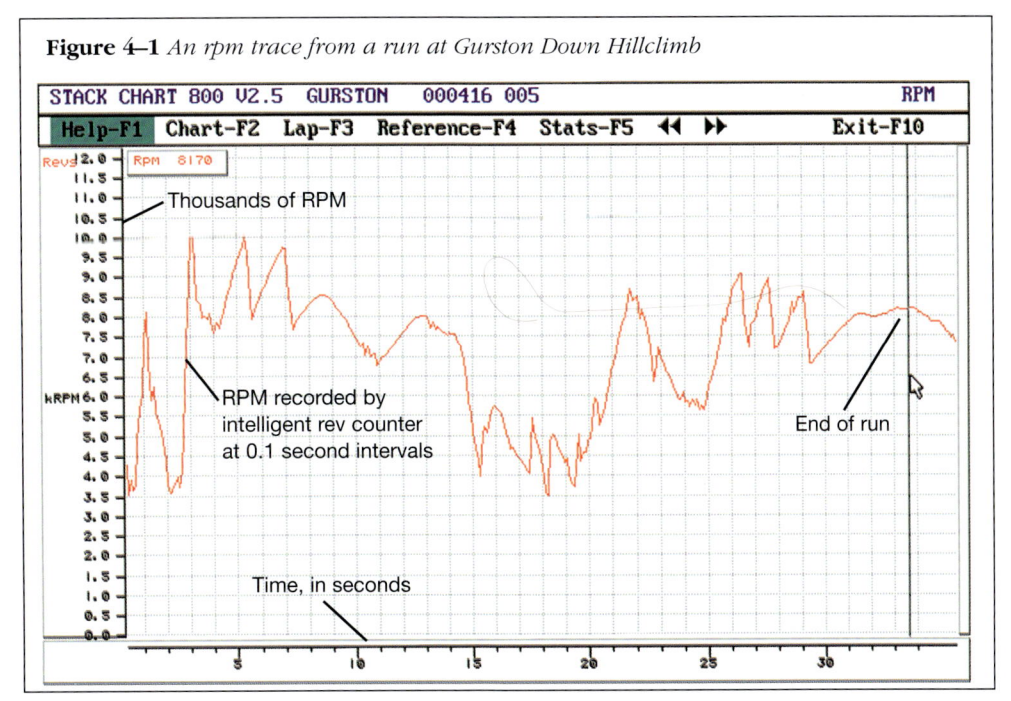

Figure 4–1 *An rpm trace from a run at Gurston Down Hillclimb*

Figure 4–2 *Matching the data to the course*

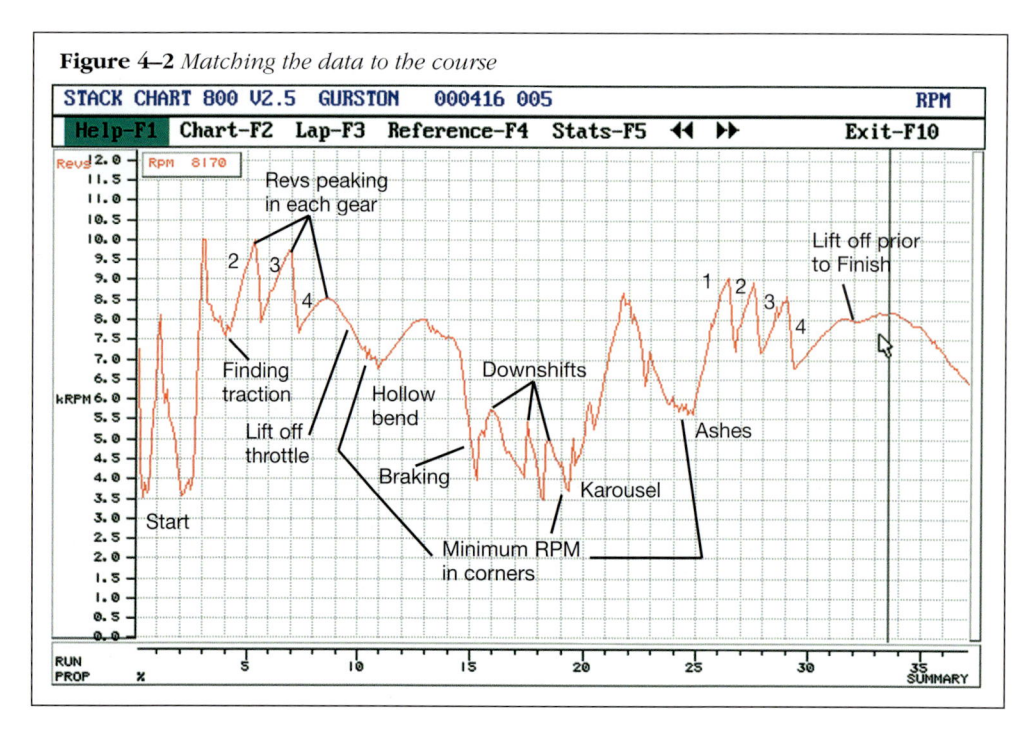

[Chart: STACK CHART 800 V2.5 GURSTON 000416 005 — RPM trace showing: Revs peaking in each gear (2, 3, 4); Finding traction; Lift off throttle; Hollow bend; Braking; Downshifts; Minimum RPM in corners; Karousel; Ashes; Lift off prior to Finish; Start]

the appropriate button on the dash a short interval before leaving the start line, and was switched off again after the finish.

Before we get into the specifics of what this plot shows, and there is some interesting detail that will serve to show how useful this simple type of plot actually is, let's make sure the basics are clear. The squiggly line shows the engine rpm during the run. The sets of peaks that look like the teeth of a coarse woodcutting saw show the revs rising as the car accelerates up through the gears, as shown in Figure 4-2. The maximum revs used in each gear can be used to work out the maximum speeds achieved in each gear, although this is of most use in the highest gear, and it represents the top speed achieved on the fastest straights. The sharp drops in rpm after each peak show the revs dropping during the gear shift as the driver selects the next gear.

Downshifts can sometimes produce confusing traces, until you get used to what you need to look for, and the telltale trace shape all depends on the car and driver technique. In the area marked 'downshifts' on Figure 4-2, the driver is using the heel-and-toe technique, using the side of his right foot to blip the throttle during braking and whilst the clutch is disengaged in order to assist the smooth transition into the next lowest gear. This has produced short rpm peaks with steep left sides indicating the rapid acceleration of the unloaded engine whilst the clutch is in, and less steep right sides showing the car's deceleration under braking with the clutch engaged. You can actually see the three downshifts from fourth gear down into first gear, and these have been labelled with the gear numbers to help pick them out. The throttle blip during the shift from fourth down into third was not so briskly performed as the remaining two shifts.

The rev trace shows drops in a number of places in addition to during gear shifts. Two general occasions are where the driver backs off the throttle to slow the car but doesn't brake, and also where he comes off the throttle and brakes as well. The latter obviously slows the car more rapidly than the former, and so the rev drop occurs more rapidly than when the driver just lifts off. Such occasions are marked on the plot in Figure 4-2.

After reaching the maximum revs attained in fourth gear in the first part of the trace, the driver lifts off the throttle to slow the car and this is revealed in the gently downward sloping part of the trace indicated. We'll go into why this was actually done here a bit later, but for now it is pointed out simply for the purpose of identifying the trace shape. The trace looks completely different when the driver applies the brakes to slow the car however, as indicated on the plot. The trace falls much more steeply in direct relation to the rate of deceleration (which you can calculate as we'll see later) until the throttle blip on the first downshift intervenes. The car continues to slow of course, because the driver is still on the brakes and coming down the gears, but the rate of deceleration cannot be seen because of the rev peaks as the throttle is blipped.

Some of the 'troughs' in the trace are where the car is going around a corner. The drops in revs before a corner show the car slowing down, the number of revs in the bottom of the trough indicates the minimum revs in a corner, and the rise in revs after a corner shows the car accelerating again. If you know the speeds in your gears you can work out the minimum revs as mph, and we'll go into that in more detail later too. Then there is an amount of detail that at first is confusing, but it can all be deciphered, some of it just by looking at the plot in isolation, whilst some of it requires a knowledge of the course layout to be fully understood. That is solved here by reproducing a map of the course as a reference (Figure 4-3). You will do this yourself with whatever form of data logging you run, either by looking up the course map in the event programme or your circuit guide, or as we shall see in the next chapter, if you get a more sophisticated data logger it may well produce a track map for you.

Matching the data to the course

The course map of this particular hillclimb, which just happens to be my local venue where I have the honour to be an instructor, is Gurston Down in Wiltshire in the beautiful chalk downs of southern England. It shows a relatively simple and short layout, and it is one that is easy to get to grips with in data interpretation at least! Illustrating with the simple and short data plot that goes with it is useful because there's plenty of basic features to see here without being swamped by data. Let's go through this run so that we can see how to match the trace to the course, and we'll look at some of the detail more closely along the way.

First it will help to know that the car being logged here was a single seater powered by a 2-litre mid-mounted engine with around 280bhp or so driving the

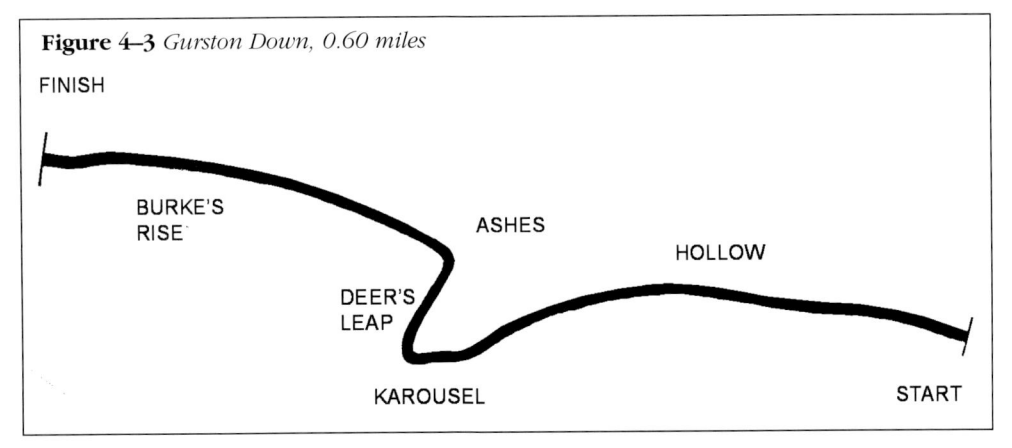

Figure 4–3 *Gurston Down, 0.60 miles*

FINISH

BURKE'S RISE

ASHES

HOLLOW

DEER'S LEAP

KAROUSEL

START

rear wheels. Thus it was more than capable of spinning the wheels quite vigorously off the startline if the driver induced that condition. Because of this, the actual instant at which the driver's start occurred is not always easy to spot precisely. But it looks at first examination as though having set the tacho into record mode the driver then blipped the throttle, and then hit the throttle hard (whilst dropping the clutch) at around the three-second point on the graph. It also helps to know that the starts were being done in second gear.

The first part of the course off the startline at Gurston is straight, and also downhill, which obviously aids the rate of acceleration that cars achieve. But because the driver is combating wheelspin, the revs rise initially and very rapidly to around 10,000rpm. It looks as though the driver's response (backing off the throttle) sees the revs drop to about 7,600rpm at about four seconds into the plot in the search for some traction. This arrives at around 4.25sec, at which point the revs rise rapidly to a maximum of 10,000rpm at 5.36sec as the car continues to accelerate.

The change into third gear is followed, not surprisingly, by the change into fourth gear, until at about 8.7sec the rev trace starts to drop. The driver has backed off the throttle here to slow the car for Hollow Bend. This is one of those corners that doesn't look much on the course map, or from the trackside, but it commands a lot of respect from the majority of the competitors. It's a fast left-hander with a slight brow on the apex which tends to unweight and destabilise the car slightly. Hence, on a good day a decent driver in a well set up 2-litre car with lots of grip (or lesser amounts of horsepower) can take the corner without slowing, but otherwise at least a lift off the throttle is required. In this case, the conditions were wintery-cold (it was mid-April and snow fell the day before) and drying from earlier rain, which helps to explain the big drop in revs as the driver came off the throttle for quite a long time. (Note: file this under 'T' for track conditions in the Driver's Book of Excuses!)

Once past the apex of Hollow Bend (at around 10.95sec) the driver is able to get back on the throttle again, hence the increase in rpm for a couple of seconds before slowing down for the Karousel hairpin complex, a two-part corner. At around the 13-second point, the driver appears to slow gradually at first, as if he'd come off the throttle but wasn't braking. This is actually where the second apex of Hollow Bend occurs, and in the prevailing conditions the driver has obviously backed off slightly to deal with this.

Then at 14.3sec the revs drop much more rapidly as the course straightens out and driver brakes hard for the first 45 right-hander of Karousel. The first downshift probably occurred before the first part of the hairpin complex, after which the track incline becomes very steep before turning sharp right through 135°. The second pair of downshifts (and additional braking) occurred before this tight part of the hairpin, which is driven in first gear. The minimum revs in this corner were just 3,720rpm at around 19.4sec, corresponding to less than 25mph, which shows how tight the corner really is.

Coming out of Karousel Hairpin, traction with this amount of horsepower in first gear once again becomes a bit of an issue, and there are two spikes in the generally increasing rev trace that show a loss of traction, either by just spinning the wheels or very likely by the car exhibiting some oversteer, that is, tail-out attitude. The spikes show the revs increasing as the wheels spin, and the mini-troughs immediately after them are the result of the driver backing off the throttle to catch or modulate the problem. At just past the 20-second point full driven-wheel grip is restored, and the car accelerates briskly over Deer Leap, peaking at just under 8,700rpm at around 21.7sec.

Here the trace becomes a little bit confusing. With a car of this power, and in spite of traction problems coming out of Karousel, the speed reached at the aptly named Deer Leap is sufficient to make the car go light enough to lose driven-wheel grip again (indeed more powerful cars can take off completely here!) so the peak of 8,700rpm is slightly false, with wheelspin masking the car's true speed. To contain the engine revs the driver has lifted off the throttle and then briefly reapplied it before braking for the important 90° left-hander called Ashes Bend. Speed carried through here helps a lot on the uphill straight that follows.

The rev trace in Ashes bottoms out at 5,640rpm or thereabouts at 24.64sec, corresponding to about 34mph. This is a little slower than on a good day and again probably relates to the track and weather conditions. But also relevant is the shape of the rev trace in this corner. Notice how, after the braking/slowing phase in which the revs drop fairly smartly, the trace shows a continued slight downward trend with small, short-duration peaks within it. This pattern may have been caused by understeer, that is, the front of the car pushing on and generally being unwilling to point around the corner the way the driver wanted it to go (we'll go into this in more detail later in this chapter). Finally he got some grip at the front, and the car ended up pointing out of the corner, and then he could reapply the throttle at around 24.9sec.

The next part of the course is uphill but fairly straight, so the trace shows the driver going up through the gears and eventually into fourth gear. But there is a tricky left–right kink over the finish line which on a good day with this type of car can be taken without lifting off the throttle. However, on an iffy day like this one, the dip in the rev trace before the finish shows the driver having a big 'confidence lift' to make sure the car stayed on the road.

As with the start, pinpointing where the car crossed the finish line with this type of logging system is not easy because the driver is obviously going to be on the throttle until after the finish line, and especially so at Gurston where the finish is on a bend and lifting off too soon and too suddenly can have dire consequences. So using a simple rev logger to time runs or laps without some type of supplementary beacon (more on those later too) is not really possible, at least not to any accuracy. But if we hazard a guess that the finish line was crossed at around the 33.5sec point, then the original notion that the start occurred at about the three-second point has to be questioned. That would equate to a 30.5sec run, which would have been a pretty competitive time for a run in good conditions, and as we have seen, this was not a run when the driver was really 'on it'.

Let's go back and analyse that start again. In fact the driver, anticipating little grip following the slippery event the previous day, dropped the clutch at the 1.14sec point with the revs set at just over 8,100rpm. However, there was actually more grip than expected, the engine 'bogged down', and the rpm dropped to under 3,600rpm at 2.10sec. This was well under the useable rev band of the engine, so after briefly, and hopelessly, pressing the throttle again (the little rise in revs up to the 2.40sec point), the driver banged in the clutch again, gave the car a bootful of angry throttle and basically restarted the run at the three-second point. That the next couple of gear shifts were made at well over the normal shift rpm of 9,500rpm on this car is also evidence of the driver's instantaneous mood!

Now that the start point is more accurately known on the trace, at around 1.14sec, and the finish was at about 33.5sec, we can surmise that the run time was 32.3 to 32.4sec, a good two to 2.5 seconds down on a decent time for the car, but not bad in the prevailing conditions with a bogged start! But establishing the run time like this is not so important as delving into the other detail to see what can be learned.

Here endeth the first lesson in data interpretation then. Already we've gone from figuring out what the squiggly line represents to interpreting what the car and driver were up to. But with each and every parameter that it is possible to log, including rpm, it pays to stop and think what influences car and driver behaviour have on that parameter. So what is rpm related to? A list would include: engine characteristics, gearing, speed, acceleration and deceleration, throttle position, traction and handling. Without actually mentioning that list we touched on each of those factors during our run up Gurston Down in the preceding section, and we also picked out some interesting details by closely examining the rev trace. But there's more . . .

Engine characteristics
Different engines in different motorsport disciplines will exhibit very different rpm traces. But being able to log the rpm gives you and your engine builder (if that's not you) hard evidence of what your type of competition is actually demanding of your engine. You can see at a glance the peak rpm at various points in a run or lap or session. Perhaps of even more value, you can see what minimum rpm your engine is having to pull from when exiting the slowest corners. These two simple facts allow you to try to match the engine's usable power band to the actual demands of competition, either by better optimising the car's gearing, of which more in a while, or by altering the specification of engine components to produce torque and power at the rpm they are needed.

It is often possible to use the software with your rpm logger to do a little bit of statistical analysis which can really help with this aspect of development. This analysis plots a bar graph or histogram of the time spent in various rpm bands. You can adjust the rev bands to those that interest you, and see very easily where your engine spends most of its on-track time. This can also be a help when analysing driver performance too, as suggested in this demo chart, but more on that later.

The rpm trace from the Gurston hillclimb illustrates that not only does a hillclimb engine need good 'top end' power to produce maximum acceleration and good straight line speed, but it must also be capable of pulling cleanly from low rpm in this particular discipline. Similar characteristics are probably required of a special stage rally engine too, but for circuit racing where speeds, and hence rpm tend to be at the higher levels for more of the time, the ability to pull from speeds as low as 25mph is almost irrelevant. Different tracks as well as different disciplines also place different demands on engines, so it's a case of horses for courses, but deciding on which type of horse is made much easier if you log some rpm data!

Clearly this sort of information provides useful fodder for discussions on tuning, and even though the occasions an engine may have to pull from low down in the rev range may be few, the impact on run or lap times may be very significant. Which just goes to prove that logging the data is one thing, but it's what you do with it that really matters!

Gearing
Most competition car transmission systems offer the ability to rapidly alter gearing by one means or another, whether it is by a change of final drive ratio, or changes to the gear ratios themselves, or maybe both. Sometimes these jobs are not practically achievable in the paddock but can be done back in the raceshop. In any event, it is usually possible to alter a car's gearing to match it to its sphere of competition, and to particular venues. It's a job that goes hand-in-hand with obtaining the right engine characteristics. If you've taken the trouble to set the engine up to match your sport and the tracks visited then you had better think about fine-tuning that match with the right gearing too!

Once again, rpm logging comes to the

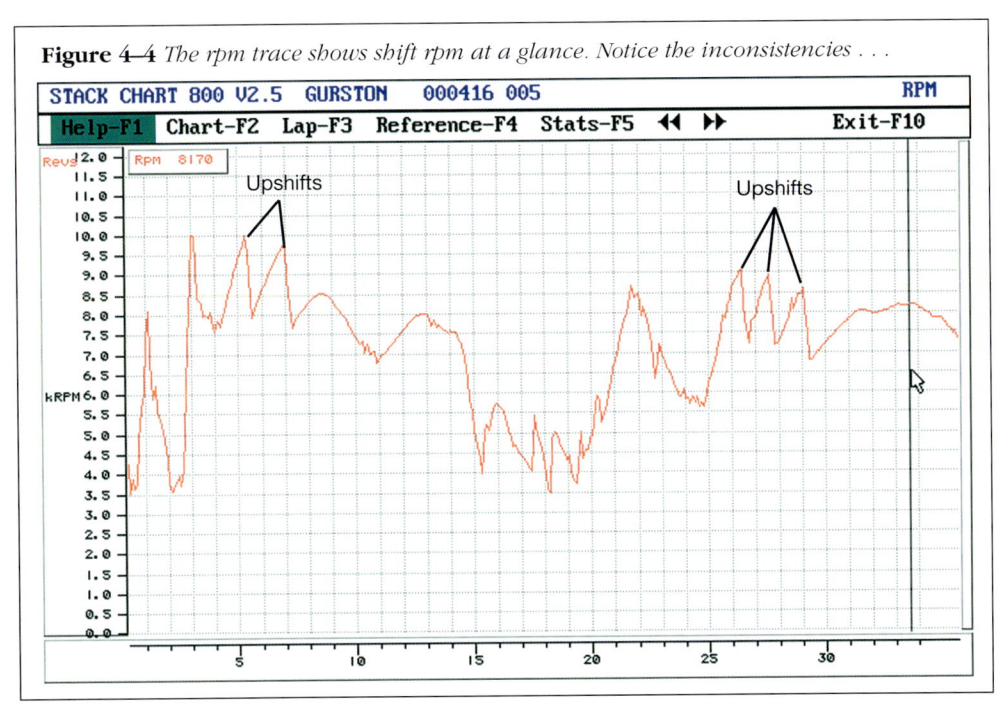

Figure 4-4 *The rpm trace shows shift rpm at a glance. Notice the inconsistencies . . .*

rescue. Without it you might be able to recall how well the engine pulled out of the corners, or whether you were running out of revs on the fastest part of the course. But you might not, and in any case there's more to it than that. Getting the best performance out of a car is not just about having the right gearing installed, but also ensuring that the driver uses the gearing to best effect. And that's something you are only really going to know if you log and study the data.

Referring once again to the rpm trace from the Gurston hillclimb (Figure 4-4, now 'zoomed in'), it shows a couple of driver inconsistencies that illustrate an important point about using the engine effectively. After his – how can I put this gently – completely bungled start, the driver made his next two gear changes at well over the normal shift rpm. The first of these was definitely excessive, although the second was maybe only a couple of hundred rpm over normal. On the second part of the run however, from Ashes Bend towards the finish, upshifts were being made at lower rpm, the first two at around 9,000rpm and the third at about 8,500rpm. In the driver's defence, the upper half of the course takes longer to dry after wet weather and this may have prompted him to 'short shift'. We'll give him the benefit of the doubt.

The ideal shift rpm obviously depends on the engine and where in the rev band it makes its best power and torque; your gearing; how the ratios step up between gears, and also how fast you are going because rolling resistance increases with speed and aerodynamic drag increases dramatically with speed. For the mathematically inclined, the aerodynamic drag force rises with the square of speed but the horsepower absorbed rises with the cube of speed.

This means that the higher gears need to be more closely spaced than the lower ones so that the revs stay further up the rpm range, and hence nearer peak power rpm, when changing into the next highest gear. It will obviously help then if you have at least the figures for peak torque and peak power and at what rpm they occur with your engine, but it is greatly preferable if you have power and torque curves to hand.

You would usually pick your gears by selecting a first gear that deals with exiting the slowest corner on the course without the rpm dropping too low, the tallest gear so that it deals with your maximum speed on the fastest straight, and the others slot in between to provide a compromise between the best acceleration and matching the intermediate gears to the other corners on the course. Now you can work out what the rev drops will be when changing up through the gears, or you can look them up on a suitable ratio chart if you have one, and relate them to your power curve.

To simplify matters, picking a fixed-shift rpm for all the gears makes it easier to remember when driving, but as mentioned two paragraphs ago, this will depend on having picked ratios that get closer together the further up the 'box you go. Best acceleration is found by looking not only at what the revs climb to in one gear, but also at what they drop to in the next gear.

Then it's a case of studying your rpm traces to ensure that you do change gear at the right rpm. Revisiting Figure 4-4, we can see the driver not making the best use of his engine. Look at the second part of the course as he accelerated towards the finish. The three gear shifts were all carried out at different rpm, and the driver was changing up at successively lower rpm as he went up the 'box. With the first two upshifts the revs dropped to about 7,200rpm but the change to fourth (top) gear saw the revs drop to about 6,800rpm, quite a bit further down the power curve.

Another aspect of shifting often overlooked, but which has a big effect on acceleration and elapsed time is the speed with which upshifts are accomplished. By running the cursor from the peak rpm at the shift point along to where the revs start to climb in the next gear, you can measure the duration of your upshifts. Take the three upshifts mentioned in the previous paragraph. These each took 0.30 seconds, which together adds up to nearly a second for which the car was not accelerating to its full potential. So upshifts are worth doing quickly, and rpm logging lets you measure how quickly you've done them!

Gearing for corners
Gearing correctly for corners can make a huge difference to accelerating away from them as quickly as possible. Pick too high a ratio and the car may actually drop out of its optimum power band, even 'off the cam' if it's a peaky engine, a problem often accompanied by breathless spluttering (from the engine and possibly the driver) if it uses what one might call an old-fashioned induction system (carburettors). Even with modern engine management systems, which enable an engine to pull smoothly under most conditions, if you fall out of the power band you will still lose out on acceleration. Pick too low a ratio however and you will spend so little time in it before changing up again that you will probably be better off not changing down that extra cog, and staying in the next tallest gear instead.

The hillclimb rev trace shows an example of a corner that does compromise the gearing slightly (see Figure 4-5). Ashes Bend, the right-hander before the last straight, is an important corner. If a car exits this too slowly it loses out all the way up the finish straight. So not only do drivers have to carry as much momentum as possible through this corner, but also they have to gear to ensure that their engines pulls strongly out of this corner. It's typical of many a corner at many a venue around the world.

The rpm trace we are looking at here shows the minimum rpm of this car in this corner to be around 5,600rpm, which is actually right at the bottom of the useful rev range for this particular engine. So the ratio has to be low enough in this gear (first) to pull cleanly away. And in this instance, the rpm trace rises swiftly once the car is lined up for corner exit, showing that the engine did pull strongly from

Figure 4–5 *Gearing for a particular corner*

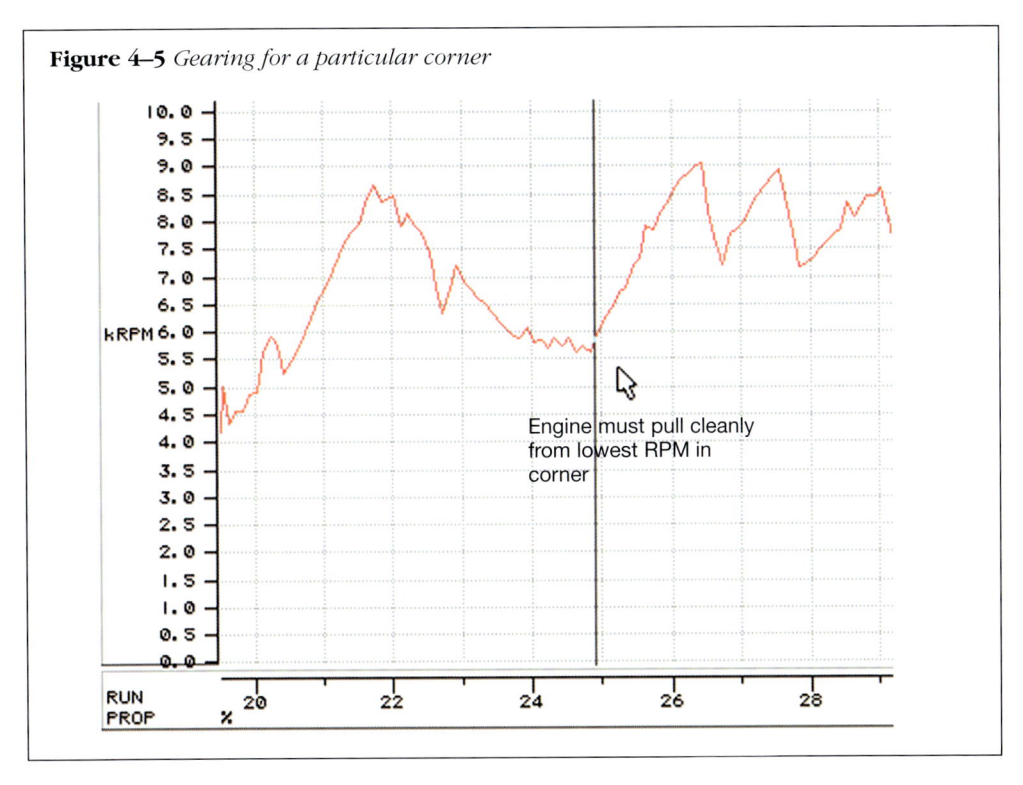

Engine must pull cleanly from lowest RPM in corner

these rpm. It's easy with just a rev trace in front of you to check out how well your car is accelerating out of corners, and to make decisions on what gearing is going to produce the best performance.

Furthermore, if your transmission allows you to alter individual ratios it is also possible to calculate what gear ratio you would need to install should you find you're running too tall a gear for example. It's easy enough with calculator to hand. If the minimum rpm in a corner as shown on your rev trace are, say, 4700 and you know that your engine only pulls strongly from 5,000rpm, then you need to drop the gear ratio by 4,700 divided by 5,000, that is 4,700/5,000 = 0.94. So if the gear was good for, say, 75mph at shift rpm, then a gear that would get you to a minimum of 5,000rpm in this hypothetical corner would actually be 0.94 x 75 = 70.5mph.

Now look up on your gear chart which ratio would give you 70.5mph at shift rpm and prepare to fit that one instead. But before you do that, think about the rpm drop to the next gear up and decide whether you will compromise the gear spacings too much. Then decide which compromise is the more important – overall acceleration through the gears or the ability to accelerate out of this particular corner (and others like it perhaps on the same track).

Gearing for maximum speed

The usual approach to gearing for maximum speed on a course is to say that the engine should be at maximum power rpm as it achieves its highest speed on a course. Thus, if your engine produces peak power at 9,500rpm, then this approach implies that you should gear it so that you pull 9,500rpm at the end of the fastest straight.

But think about the shape of your power curve for a moment. Although it does tail off more or less rapidly, power doesn't just disappear once you passed peak power rpm. The 280bhp engine in the hillclimb car example in this chapter still perhaps makes 270bhp a few hundred rpm past the power peak of 9,500rpm. If you were to gear the car so that its maximum speed occurred, say, 300rpm past the peak power revs with a slightly lower ratio, then you would gain on two fronts. First, the slightly lower ratio would accelerate you a little bit more rapidly, and secondly, the revs you drop to when changing up into top gear would be higher, putting you further up the power curve, and this again would give you more acceleration. It might be a good idea to discuss this strategy with your engine builder though. He might be less than sympathetic if you don't ask him and then your engine goes bang.

Making this sort of change and then studying the comparative data on your rev trace will show these points. The revs you drop to as you change into top will show as being higher, and the gradient of the rpm trace as you build speed in top gear will be steeper than previously, indicating more rapid acceleration. You could expect to see a slightly higher maximum rpm on the straight too, all other factors being equal, unless you are nearing the aerodynamic limitation on top speed where horsepower absorbed by drag (aero and mechanical combined) equals horsepower being produced at the wheels. That would be evident as a flattening out, ultimately to horizontal, of the rev trace.

We seem to have directed plenty of time to discussing gearing here, but it always surprises me how little attention some competitors pay to the topic. Getting the gearing decisions right can often be a better value-for-money way to improve performance than many others, and while you don't necessarily need a rev logger to come to these decisions, they're a whole lot easier to make if you have rpm data in front of you.

Calculating speeds

Engine rpm equates directly to speed except when the driven wheels are spinning or locked up during braking. To work out the speed at any particular point you need to know what gear the car is in and what the mph per 1,000rpm in each gear are, and by reading off the rpm at points of interest you can then work this out as speed in mph. It's easy enough to figure out what gear the car is in from the rev trace providing you know how many gears are being used, and the calculation is pretty simple.

It isn't actually necessary to calculate speeds but it sometimes makes comparisons easier, especially if you are in the happy position of being able to compare speeds with different drivers in different cars running different gearing. As an amateur competitor I can honestly say that this situation has usually been the norm in my sphere of competition, although no doubt any serious and/or professional racers who might read this will consider it either to be fantasy or at the very least the giving away of valuable information! But even if you cannot make these sorts of comparisons, if you alter your own car's gearing it will make before-and-after comparisons more meaningful if, for example, you assess changes in end of straight speeds in mph rather than rpm. Similarly corner speed comparisons are more easily compared in mph than rpm.

Calculating gearing and putting your available gears into meaningful mph per 1000rpm figures is easily done. You need some information first, though; you need to measure the driven wheel circumference (marked off on the floor at correct tyre inflation pressure and representative car weight); and you need to know the final drive and gear ratios. To calculate the speed in each gear at any rpm you work out the following:

mph/1,000rpm =

$$\frac{\text{driven wheel circumference (feet)} \times 60 \times \text{rpm}}{5,280 \times \text{gear ratio} \times \text{final drive ratio}}$$

k/mh/1,000rpm =

$$\frac{\text{driven wheel circumference (metres)} \times 60 \times \text{rpm}}{1,000 \times \text{gear ratio} \times \text{final drive ratio}}$$

where the gear ratio is expressed in the format '1.8' where it is a '1.8:1' gear. A gear ratio referred to as 18:32 would need to be calculated as a 32 divided by 18 or 1.777:1, and the number 1.777 entered into this formula. Similarly, a final drive ratio expressed as 8:35 would be calculated as 35 divided by 8 or 4.375:1, and 4.375 entered into the equation.

You can work out a simplified formula for your tyre sizes and final drive ratio, and once you've done that it's very quick to look up the rpm on your rev trace and calculate the speeds in each gear at any point of interest on a course. You can then also start an objective analysis of the proportion of time spent around a given track in different speed ranges, and this can form a part of more scientific assessment of your car's aerodynamic requirements. You might be thinking that going to this depth would be more easily achieved if you logged speed rather than calculating it. Of course it would. It would also be more expensive than just logging rpm and doing a bit of leg work with a calculator. But it goes to show that you can derive masses of useful objective data if you are prepared to study your rpm trace and squeeze every last drop of information out of it.

Some suppliers of rev loggers offer features built into their analysis software that enable you to enter your speeds in each gear, and then calculate from the rpm at any point on the trace what the speed is. This obviously saves a bit of time, but the end result is no different to tapping a few numbers into your calculator.

Analysing the driver

We have already picked out a few points from the rev traces that relate to driver technique, for example the messed-up start and the inconsistent gear change points on the run illustrated at the Gurston Down hillclimb. These are important applications of rpm trace analysis in any discipline, startline technique perhaps being thought of as especially critical on something like a short duration hillclimb run — but in what motorsport discipline is startline technique unimportant? It's just as vital to get the start right on the race track, and it's probably the most important factor in drag racing after driver reaction time, or so they tell me.

Proper use of the performance that the engine can deliver is crucial in any discipline too. I recall once being gently lambasted by one engine builder for short-shifting on a hillclimb event on a part of the course where it felt more 'secure' to do so . . . I wouldn't mind but he wasn't even the guy who built our engine! Perhaps it was for this reason (and the fact that he was right) that his point stuck with me, and it's something that's easily checked with an rpm logger so that you can modify driving technique if needs be.

Another aspect of technique that can be checked easily is the revs seen on downshifts. There are two things to look for here. The first is the revs seen during the heel-and-toe throttle blip, when the clutch is in and the car is effectively in neutral. The engine is off-load in this condition, and over-enthusiastic or clumsy throttle blipping can see the engine revs climb way too high. According to no less an authority than Carroll Smith, this is the most potentially harmful form of over-rev, so is probably best avoided! The telltale peak in the rpm trace will be very evident.

The second, related occurrence is caused by the driver changing down too early in the sequence of slowing for a corner to the extent that the revs in the next lowest gear as it initially engages, exceed safe limits. It happens when the driver changes down at the same time as he begins his braking, instead of a short time after that point, so that little speed has been removed, and the speed is just too high for the lower gear. This causes an over-rev which no electronic rev limiter can do anything about — the car's speed just mechanically over-revs the engine. Mr Smith reckons this to be the second most harmful form of over-rev. Again the occurrence is easy to spot on the rev trace. It will be distinct from the above-mentioned off-load over-rev because of a less steep left side to the peak. Off-load the engine accelerates extremely quickly, and a very steep left side to the peak is created.

It is also possible to look at the driver's technique exiting traction-limited corners, such as slow hairpins where wheelspin is easily induced. There are two possibly unrelated issues here though, and the hillclimb example again serves to illustrate this. Look at the middle part of the run again (Figure 4-6), where the car is in the slowest corner (Karousel), and the revs drop to around 3700 in first gear at the 19.3-second point. As the car exits that corner, there are two spikes in the engine's acceleration curve, the first at 19.5sec and the second at 20.2sec. These were caused by the driver modulating the throttle to contain wheelspin and/or oversteer, as was stated earlier. The wheelspin may have been related to the car oversteering (it being a rear-wheel-drive car – the opposite may be true if it were front-wheel drive), and although it is not possible to determine this conclusively from an rpm graph in isolation, a trace like this serves as a useful memo to ask the driver, or for the driver to remind him or herself, to try to recall if oversteer did occur here.

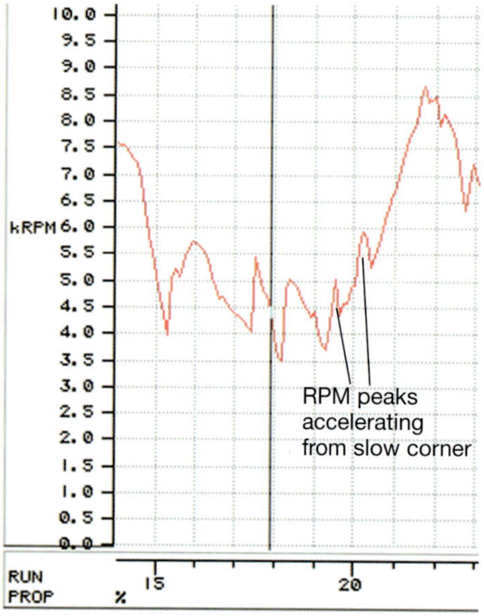

A couple of things come to mind when you spot this kind of occurrence. First, if the car did oversteer, maybe there is something you can adjust to improve things, and more on that shortly. But, in terms of driver technique, is the driver simply stepping on the loud pedal too hard, too soon? A car that is spinning its wheels to the extent that the driver has to back off the throttle twice in succession is scrabbling for forward traction. If the driver has induced oversteer too by being over-aggressive on the throttle, then the car will have reduced forward traction. So, this type of trace should make you ask if the driver can smooth things out a bit with more progressive throttle application, because the chances are that the car will come off the corner more effectively with a slightly smoother touch.

The same type of analysis can also be made on the rev trace of the driver's start, and on other areas of the course where straight line traction is an issue. You have to keep an open mind though to decide whether the chassis is the limiting factor, and if so, what if anything can be done to improve things. But if, in the short (or the long) term there is nothing to be done, then a modification of driver technique may be the best way to maximise available acceleration. Naturally, and rightly, the driver should not be content with this state of affairs, and neither should his race engineer if he has one, because there may be technical improvements that may help the problem. But sometimes short term palliatives are the only available, practical option.

Analysing aspects of chassis performance

Armed with just an rpm trace you can also begin to analyse what your chassis is doing, but there are limitations here which we'll discuss as we go along. Some

Figure 4–6 *Loss of traction exiting a slow corner*

of the analysis is based on solid data, and some is based on using the data as a reminder, as discussed in the previous section.

Longitudinal G

One of the simplest things you can do is to calculate your car's rates of acceleration and deceleration, otherwise known as longitudinal G. This is done by measuring on the rpm trace the change in speeds over a given time, which by definition is what acceleration is, and then calculating what this represents in terms of G. The rpm graph of the Gurston hillclimb once again serves to illustrate this (Figure 4-7).

After the point where the car begins to accelerate linearly after the fluffed start, at 4.30sec the rpm figure is 7,890. At the point the gear change was made the rpm peaked at 9580, at 5.40sec. That's all the information we need to calculate the acceleration G over this time period of 1.10sec, and the actual mathematical formula is very simple. With speeds in mph it is;

$$\text{acceleration, in G} = \frac{\text{final speed} - \text{initial speed}}{\text{Time} \times 21.97}$$

The gear in use here was second, good for 70mph at 9,500rpm. So the initial speed at 7,890rpm would have been (7,890/9,500) x 70 = 58.14mph, and the final speed was 9,580/9,500 x 31.26 = 70.59mph. The duration of the acceleration we are looking at was 5.40 - 4.30sec = 1.10sec, and putting these into the formula for acceleration in G we have:

$$\frac{70.59 - 58.14}{1.10 \times 21.97} = 0.52\text{G}$$

A bit later in the acceleration sequence after the start, when the driver changes into fourth gear, the acceleration 'slope', that is, the steepness of the rising rpm slope, is lower, indicating lower longitudinal G. We can put the numbers through

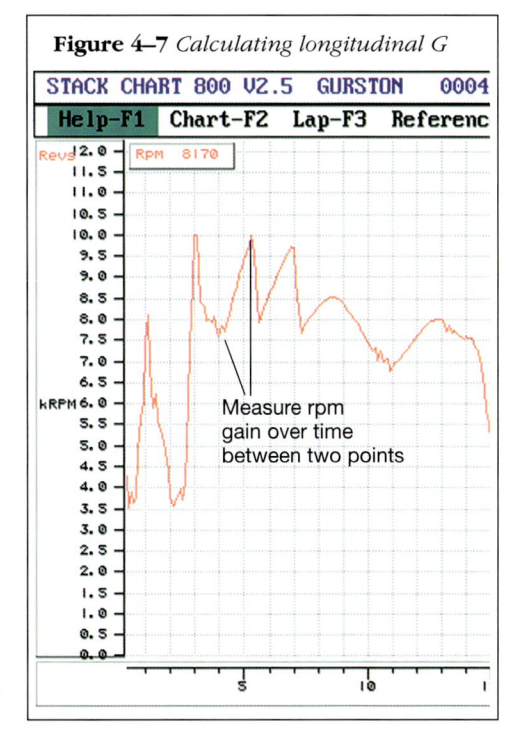

Figure 4–7 *Calculating longitudinal G*

the same calculation to confirm this. The gear was good for 114mph at 9,500rpm, and the revs rose from 7,993rpm (95.92mph) as fourth was engaged at 7.67sec to 8,525rpm (102.30mph) at 8.57sec. Thus the acceleration works out at:

$$\frac{102.3 - 95.92}{0.90 \times 21.97} = 0.32\text{G}$$

This confirms the diagnosis by eyeballing the graph that, as expected, the acceleration in fourth gear is lower than the acceleration in second gear. The actual numbers obviously depend on the gear ratios, the track, wheelspin, mechanical friction, aerodynamic drag, power and torque, tyre grip and so on, to name but some of the factors involved.

Deceleration can be calculated in the same way, using the same formula. Move along the graph a little, to the 14.66sec point when the driver is obviously slowing rapidly before changing down to the next lowest gear. The revs (still in fourth gear) drop from 6,867rpm (82.40mph) to 3,991rpm (47.89mph) at 15.36sec, or in 0.70sec. Deceleration works out to:

$$\frac{47.89 - 82.40}{0.70 \times 21.97} = -2.24G$$

Notice that the figure has gone negative because the final speed is slower than the initial speed. Notice also that the magnitude of deceleration is considerably greater than that under acceleration. With a normal two-wheeled drive car this is to be expected because all four wheels are applying deceleration effort on to the road, as opposed to just two wheels when accelerating. This type of car might be expected to achieve a peak acceleration of a little over 1G, and indeed the brief section from 20.55sec to 20.85sec, where the rpm curve rises very rapidly in first gear, shows an acceleration from 37.56mph to 44.34mph in 0.3sec, corresponding to 1.03G.

Why is there this limitation? Because that's the best that the tyres in use are likely to give, and this is where G analysis comes in useful. You can check with your tyre supplier to see what sort of friction coefficients your tyres are likely to offer, and this will enable you to work out roughly what levels of G you should see. In the absence of downforce your tyres' friction coefficient will approximately equate to your acceleration in G when all four wheels are contributing.

Thus, tyres offering a coefficient of around 1.6 will be capable of generating roughly 1.6G in braking, or acceleration if all four wheels are driving. If only two wheels are driving then under acceleration this will ratio down in proportion to the weight distribution so that, if say 65 per cent of the car's weight is on the driven wheels, expect around 65 per cent of 1.6G, or 1.04G as a maximum acceleration G. Downforce complicates these rough sums because it adds to the grip at higher speeds, and in a non-linear way. You can calculate it if you wish (see Chapter 2 of my earlier book *Competition Car Downforce*), but if you need to get into in this kind of depth you'll probably be looking at a more flexible and powerful data logger that logs these parameters. Suffice to say that you can at least see if you're using the brakes to reasonable effect by calculating deceleration rates, and you can also look at snapshots of your linear acceleration under power too.

From these examples it can be seen that it is possible to work out short term accelerations and decelerations where there is a more or less straight sloping section of rpm trace to work with. This particular rpm logger only samples every 0.1sec, so you really need to look at perhaps half-second sections as a minimum. Furthermore, over longer time periods the rpm slope may not be straight, especially in the higher gears, so if you sample too long a period then you're only getting a rough average acceleration. Stick to slopes that are more or less straight though and the figures can be useful.

Cornering behaviour

Many people would probably consider the analysis of a car's handling using just an rpm trace to be a somewhat over-optimistic extrapolation of the available data, and they might be right. However, it is possible to use rpm traces as memory joggers on your car's behaviour in corners, and as it happens there are two differing examples on the trace from the Gurston hillclimb we have been looking at, and we've already mentioned them in passing. Actually, it's amazing how much has been gleaned from this one trace of such a short track. It just goes to prove the point, that there is much information that can be extracted from a simple log of rpm if you spend the time looking for it.

We have already looked at the short section under hard acceleration in first gear out of a tight hairpin bend where there are two short duration peaks and troughs. Pretty obviously they have to be the result of the driver backing off the throttle and reapplying it again. Again,

Catching a tail-slide may show as spikes and dips on the rpm trace, unless the driver keeps his right foot hard on the throttle! (Tracey Inglis)

there are just two likely reasons. One, he is trying to find some traction, and two, he is trying to catch a rear-end tail slide. The relatively long duration between the two small peaks of 0.70sec is probably longer than even an amateur race driver would take to react to the rear wheels spinning through lack of traction, so that what's probably happening here is that the car is oversteering and the driver is reacting to catch this, a rather slower event.

Let's move on to the next corner, at around 23.8 to 24.8sec on the trace (Figure 4-8). This corner is not as tight as the previous one, but it does have a tendency to contribute to very different handling characteristics. The rev trace here is very different too. What is happening here? Well, the driver has slowed down for the corner, braking from 23.03sec to 23.83sec, as shown by the rapid rpm drop. We cannot be sure if he starts his turn-in before he finishes braking or waits until after the maximum deceleration has ended, but it's a fair assumption that he didn't put a lot of cornering load on until after the maximum decelerative G was completed – in fact his tyres wouldn't have let him. So, his turn-in probably started at the 23.83sec point. Just after this is a very small rev peak, and then a couple of other small peaks in what is otherwise a slight downward rev slope to the point where the minimum revs of 5,616rpm occurred at 24.82sec. The revs then rise sharply, pretty well at maximum acceleration from here going by the slope of the rev rise, as the driver powers the car out of the corner again.

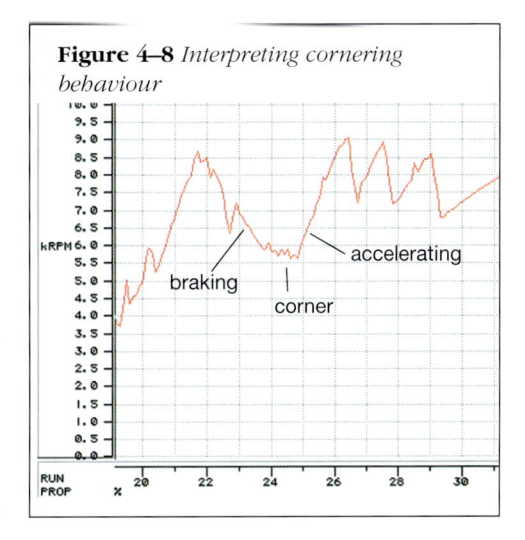

Figure 4–8 *Interpreting cornering behaviour*

But what was going on in the corner itself? Were those small rev peaks significant? Was the time delay between the end of the braking period to the point where the power was fully applied again also significant? The answer to that last point might help to explain what was going on. The duration for which the car couldn't be accelerated was almost exactly a full second, a fairly long time for this short corner. Those small peaks in the almost flat rev trough may point to the driver trying to get on the throttle but being unable to do so. The trace shape is obviously not the same as the previous corner where the driver had to back off the throttle because in that instance the back of the car broke away.

The answer then is probably the occurrence of understeer, and the credibility of this diagnosis is helped when the shape of the road here is described, because it falls away with an adverse camber part way around the corner. So the car is probably pushing on through from the middle part of the corner towards the outside of the track, and the driver is having to hold off the throttle until the car slows enough to gain some front-end grip. Once that grip, and the desired direction of travel are re-established, the driver accelerates away again.

There is another way of looking at this situation too. Maybe the driver induced the understeer by trying to get on the throttle too soon. Or maybe he just turned in too fast . . .

So these are just two examples of how an rpm trace can be used as possible evidence of oversteer or understeer. You can't prove the occurrence of either with just rpm, but the circumstantial evidence the trace provides can lead you to suspect one or the other and to ask the driver (yourself if that's you) if they remember them occurring. So if you think what the ideal rpm trace shape would be in relation to any given corner, and the actual trace deviates from it in an unusual way, ask the questions!

Data comparisons

Perhaps the most useful task you can do with any form of data logging is to perform comparisons, either with different laps or runs by one driver, or between one or more drivers in the same car, or indeed between one or more drivers in different cars. Comparisons provide you with a yard stick by which to assess performance or technique. The yard stick may not be the best or the fastest to judge against, but you do learn a whole lot more if you can make comparisons, especially with others, than if you're just out there in the wilderness plugging away on your own.

But even making comparisons with data traces from your own previous runs can be invaluable. By just using rpm logging, in no time at all you will compile a mass of data on corner minimum rpm (speeds), sector times (between rpm 'troughs' in corners), maximum rpm (speeds) between corners, and best theoretical run or lap times (calculated by adding together the best sectors of various runs or laps) for each venue you visit. This is where the value of historical data begins to show, because it's where you realise not only how you can improve performance, but also where and possibly by how much.

Comparing data with somebody in another car can have obvious value for each driver, providing openness, honesty and objectivity can be applied. This is no territory for the paranoid to tread! However, it has to be borne in mind that car set-ups may be different, and so some of the specific data will differ. Nevertheless, if you're prepared to swap corner and straight speeds then there will be scope for identifying where improvements can be made, quite probably by each driver. Of course, achieving those improvements is entirely down to the individual drivers.

The same principle applies to drivers sharing a car, although the comment above about differing set-ups obviously

Comparing data with other competitors doesn't happen in every category . . . (Tracey Inglis)

does not apply. This actually makes comparisons easier and quicker because conversions from rpm values will not be necessary. Also, your analysis software may allow you to put two traces on display at once, possibly even on the same graph so that they 'overlay' each other. This method of comparing data is probably the most useful data analysis technique you'll use, whether it's looking at just your own data or comparing it with somebody else's, because differences between runs, laps or drivers become immediately obvious. If your rpm trace falls to lower levels in corners than your car-sharing partner's, it will show very clearly on the graphs, and there will be no mistaking the fact that you are going slower than he or she is at those points.

Summary

We've spent a long time looking at rpm logging, but along the way we've also seen some of the general principles of graph interpretation and data analysis. I hope it will also be apparent just how much really useful information you can glean from an rpm trace. This is not to denigrate the value of multi-channel systems – far from it, as the next few chapters will hopefully demonstrate. But if your budget does not stretch to a multi-channel DAS, an rpm logger can supply a lot of valuable data. Professional race engineers will tell you that, before the arrival of data logging, the best drivers used to be able to recall rpm in corners, corner exits and at the ends of straights. But I bet even the best couldn't do it as well as an rpm logger!

Chapter 5

More channels!

WHILE THERE'S A lot to be gained from rpm analysis, it has to be said that the ability to log more channels can be a great help to the data-hungry driver/engineer. There is however the very real danger of data-overload, especially for the weekend amateur (most of us in other words . . .) who in practical terms will probably need to spend most of their all-too-scarce spare time at and between events making sure the car is fully and properly prepared. This means that when you've decided to invest in a multi-channel logger, the selection of which channels to log is important so that you can gain maximum benefit as efficiently as possible.

But which channels?

During my researches for this book I had the pleasure and privilege of speaking with numerous expert users of data logging, including Nigel Greensall, former UK BOSS Champion in a Formula 1 Tyrrell, and, sparing his blushes, a heck of a driver in any racecar he sits in. He is also a professional driver coach, and relies heavily on data logging to help him in his job. When asked which channels he would invest in if he just had the budget for a six-channel logger he replied immediately, 'speed and rpm'. Yes, that's only two, and I know Nigel can count. The fact is, as he explains here, he firmly believes that these two channels can provide the majority of what most drivers need to know: 'I'd always go for speed and rpm as the first two, purely because as a driver training aid it's easy for people to get a picture in their own mind of where they are on the circuit, and what they're doing (using speed). And quite often in terms of driver training it's the basics like gear changing where a lot of people are losing a lot of lap time, and looking at comparative rpm and speed traces you can tell exactly how they're doing it, and how much speed they're losing.'

This is interesting, and had I taken Nigel literally it would have made this book a lot thinner! We've already seen how much benefit can be gained from rpm analysis, and some of that analysis involved the use of a calculator to convert rpm to speed. So if we can have a system that does that for us by logging speed in the first place, then we're well on the way to gleaning a lot of the most useful information relatively easily. So, picking the first two channels was very easy then: speed and rpm.

But now to ignore Nigel's basic assertion and pick four more channels anyway, because we're interested in more! What additional channels should we select to use up our budget? Nigel does have some thoughts on this: 'Throttle position is probably the third best (choice). Lots of people talk about going flat through corners and, you know, the throttle . . . isn't completely flat, let's say, through the corner. But also some people are too aggressive on the throttle, maybe too soon in the corner. Maybe they've lost too much speed on

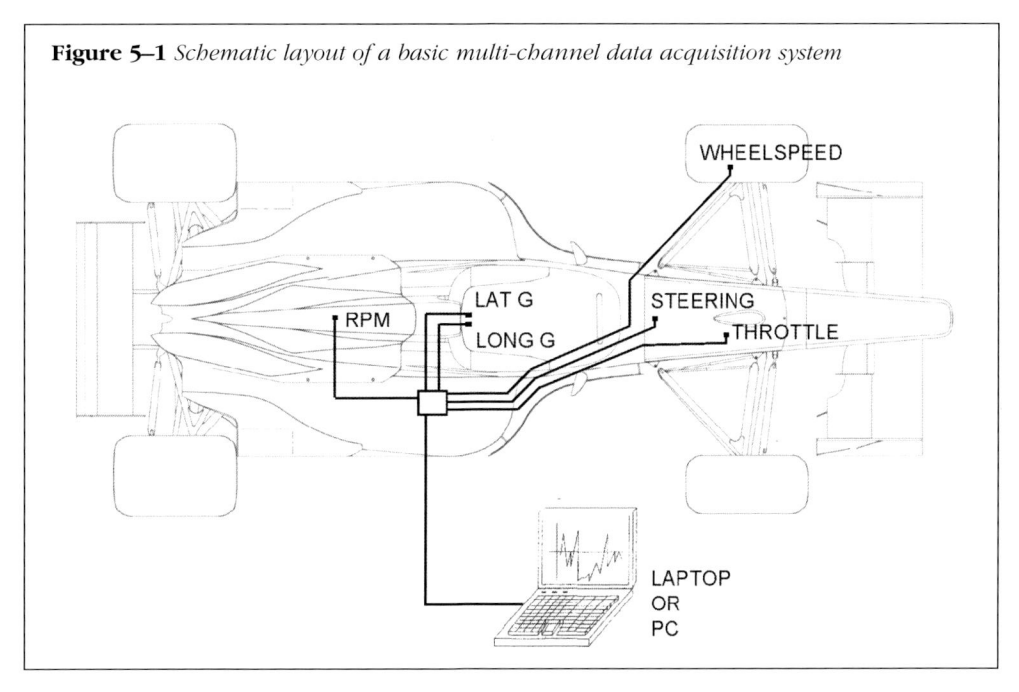

Figure 5-1 *Schematic layout of a basic multi-channel data acquisition system*

(Labels: RPM, LAT G, LONG G, WHEELSPEED, STEERING, THROTTLE, LAPTOP OR PC)

the entry to the corner, and then they use too much power too soon. That can create understeer on a car and that can cause problems, so if you can see the throttle movement, that's very useful.'

So there are more than just the obvious uses to which a throttle position sensor can be put. I once described this sensor as 'the ultimate wimp meter', a somewhat less diplomatic way of articulating what Nigel said! But logging the actual position of the throttle will tell the driver whether his belief (which may well be honestly held) that he was 'flat through Buttock Clencher Curve' was correct or not. Compulsive liars will be found out too! But as Nigel mentions, there are less intrusive results to be had from studying a driver's use of the throttle, whether he's brave in the fast corners or not. Studying the driver's use of the throttle can go a long way to explaining what might otherwise be thought of as handling imbalances . . .

So that's three out of six. The fourth channel will probably pick itself if the data analysis software can provide 'track mapping', because it is necessary to log lateral G-force to enable this function. Track mapping is performed by the software, using calculations based on the detected cornering forces combined with recorded speed over time, to produce a display of the path followed by the car. The result is an on-screen plot of this line, which inevitably mirrors the track layout. Of course, lateral G also fundamentally shows what cornering forces are generated by the car, and pretty clearly this is related to the performance of the car and the driver. So it's a very useful channel in its own right.

Then it's very possible that you would pick steering angle and longitudinal G as the next two sensors you'd buy. Steering can be informative of driver style, as Nigel Greensall remarks: 'I find steering a good one (to log). Sometimes people get very busy with the steering wheel, and they need to slow the steering down, almost steer in slow motion.' A steering trace can also be used to help pinpoint handling problems like understeer and oversteer.

Longitudinal G helps most obviously with working out braking performance but can also be used to analyse acceleration. The choice is yours of course, and you may be more interested in certain engine parameters, such as fluid temperatures or pressures. Figure 5-1 schematically shows a typical six-channel data logging system.

If your logger has a limit to the number of channels it can log at any one time but you can afford more sensors than that, and circuit racing is your game, you could always record one set of channels through testing, practice and qualifying and a different or modified set during the race. Testing and practice logging might concentrate on those channels that relate to maximum performance, including such as speed, throttle and G-forces (cornering and braking performance), whilst race logging might more usefully relate to channels that help to build an engine history (oil pressure, water temperature). The assumption here is that you have the time to analyse test or practice data between runs or sessions so you can put into practice what the data teaches you. In a race situation you probably won't be able to react to the data until the next visit to the venue, so performance-related channels are less useful. This idea of connecting those sensors that are of particular interest at any one time could obviously be applied to any discipline – you might need to study chassis-based parameters early in the season and switch to engine-based channels later on, for example.

There is another piece of equipment that will be essential for meaningful logging too – the lap beacon, which should normally come with a basic logging kit. This consists of an infrared transmitter beacon that sits trackside atop a tripod, pointing out towards the track from a location by the pit wall usually, and an on-car receiver that picks up a coded reference signal each time the car passes the beacon. This tells the analysis software that the car is passing the start/finish line, in effect, although clearly not everyone will be able to site their beacons exactly at the start/finish line! It is useful though, wherever you locate your beacon, if you put it as close to the same spot as possible every time you visit a given venue. This makes comparisons between data from visits on different days much easier because when you overlay the data, the traces will align on-screen. If the beacon is at a different point on each visit, the software will display the data out of synchrony, and it will appear as if the corners and other reference points have moved relative to each other between visits! It's not a big problem, but it makes life easier if you can avoid it.

For non-circuit logging such as on hill-climbs, it is sometimes possible to locate a beacon near the finish line, although again, not everybody who might wish to do this will be able to. Some of the data analysis software packages can cater for the non-circuit nature of these types of courses by calculating the track map from markers placed in the data, so the map is constructed between specific points in the data. This might require the driver to manually push a steering wheel-mounted button as he crosses the finish line to put in the marker that the beacon would otherwise enter. The start, which is from rest in these types of events, would be from the first instant that the logger recorded a speed value (clearly a fraction after the start, but unlikely to affect the analysis of data significantly).

More basic interpretation

By now graphs of data traces will be getting increasingly familiar, but let's just pause again to make sure we know what

An infrared lap beacon.

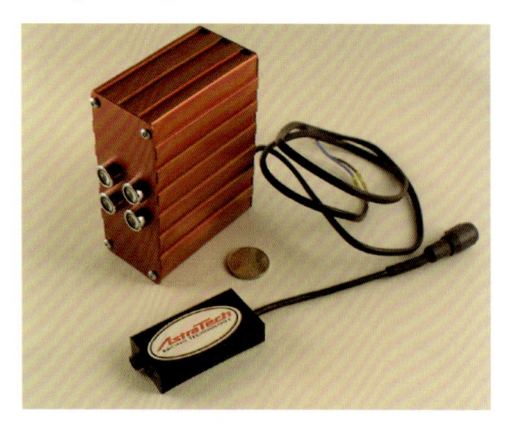

MORE CHANNELS! 57

Professional driver coach Nigel Greensall at work in an F1 Tyrrell.

we're looking at. There's more and more information to be digested in all the graphs we're to look at from now on, so the basics are important.

Figure 5-2 shows speed and rpm traces for a lap around Thruxton circuit in Hampshire, England. The car was the superbly prepared BMW M3 E36 of Martyn

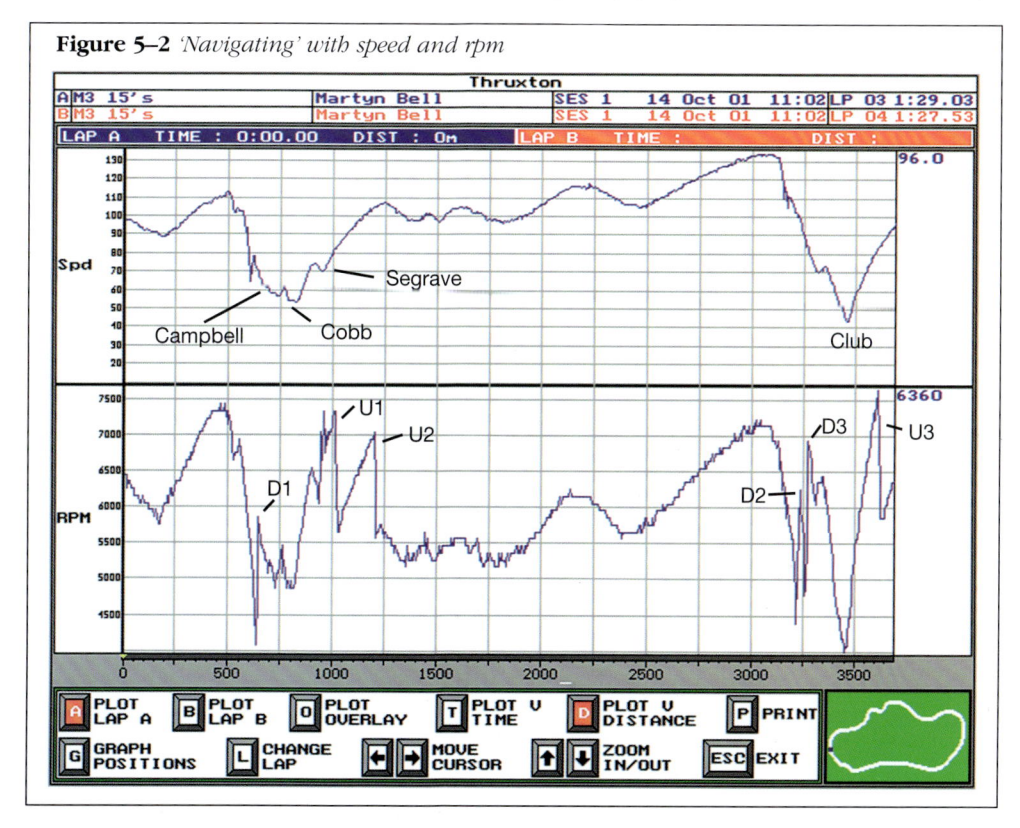

Figure 5–2 *'Navigating' with speed and rpm*

Figure 5–3 *Thruxton circuit, 2.356 miles*

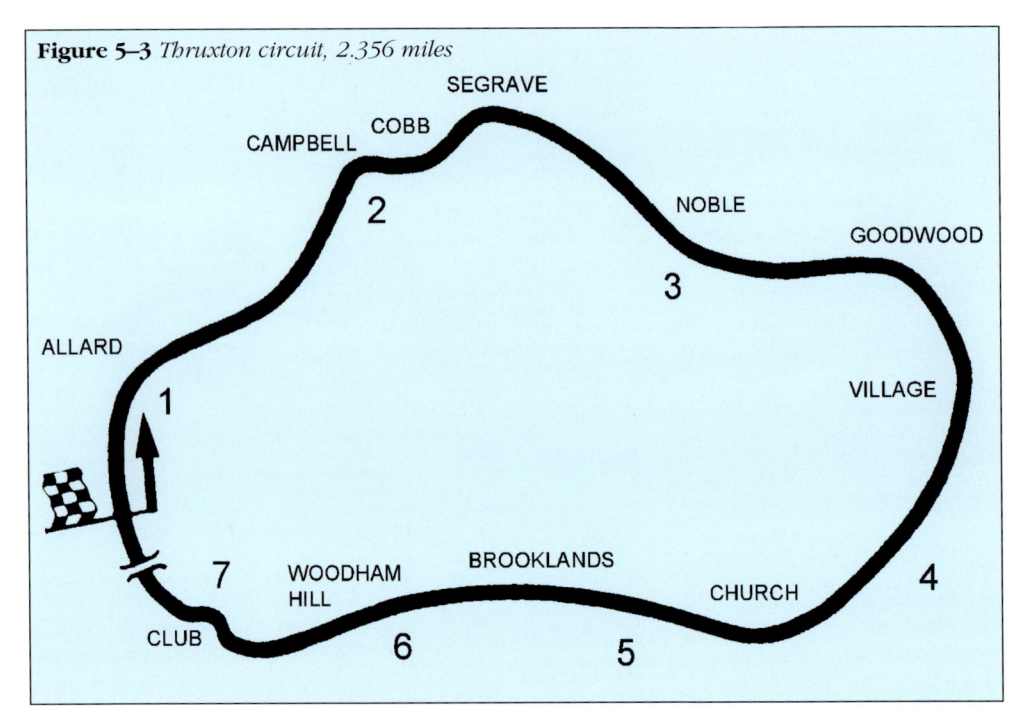

Bell, Class A Champion in the 2001 Kumho BMW Championship, and the data logging system was from Astratech. Thruxton is a fairly simple circuit from the viewpoint of working out where on the track the data refers (see Figure 5-3 for a real circuit map), with two slow sections that stand out easily on the speed trace, the Campbell-Cobb-Segrave 'Complex' and the Club Chicane, and the rest of the track done in top (fifth) gear (at least, on the lap shown here). This makes the rpm trace simple too, with few gear-shifts; there is a down-shift for the Complex,

Martyn Bell, seen here in his BMW M3 E36 in the rain at Thruxton, is an instructor at Cadwell Park race circuit in eastern England.

marked D1 on the rpm trace, into third gear; two up-shifts after the Complex, marked U1 and U2 into fourth and then fifth gear; two downshifts for the Club Chicane, marked D2 and D3 into third gear again; and finally an up-shift into fourth gear after Club Chicane.

Looking at the baseline, or x-axis of the graph now though, you can see that it's not calibrated in intervals of time, but is in fact showing distance in metres, from zero where the lap beacon is located, near the start/finish line, around to an indicated 3,700 metres, corresponding to 2.3 miles, where the official lap distance is 2.356 miles. The difference of 2 per cent will be due in part to the line taken by the car, as opposed to the distance around the surveyed 'line', and probably a small error in the measurement of tyre circumference when calibrating the speed channel.

By using distance on the baseline, all the geographical reference points, that is, the corners and straights, will be plotted at pretty much the same point on the graph lap after lap (so long as the driver's line is reasonably consistent anyway . . .). If time was used on the baseline, these reference points would relate to when the car reached them rather than where they were on the track. As lap times vary from lap to lap this, for example, this would put minimum speed 'troughs' in the speed trace in different places on the graphs if the driver arrived at these points at different times.

Zooming in is possible, as shown in Figure 5-4, which is a section of the same trace with the cursor (the vertical black line) centred on the apex of Cobb Corner in the middle of the Complex. You can see how this is useful for looking at the detail more closely. For example, just to the right of the cursor at the 950-metre point is a section of rpm trace that looks interesting, coinciding with what may be some lively chassis or driver behaviour! We'll come back to that later . . .

We have discussed previously the benefit of comparing data, either run with run, or lap with lap by one driver, or between

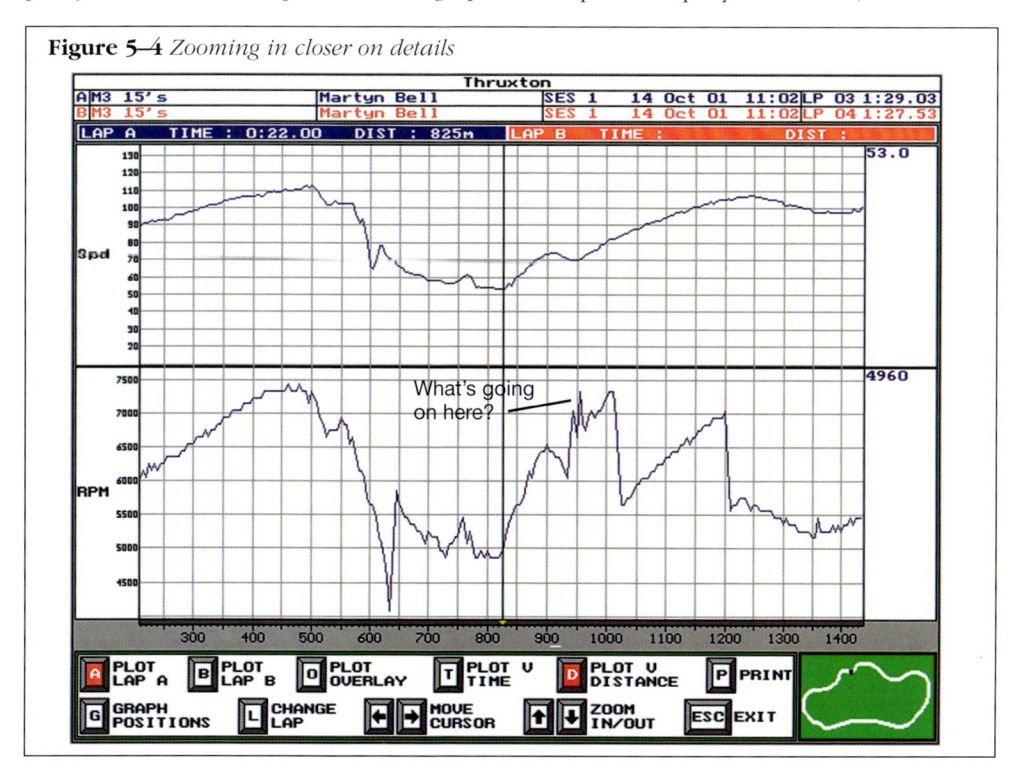

Figure 5–4 *Zooming in closer on details*

drivers. The easiest way to do this is with data 'overlay', and the Astratech software, in common with just about every other data analysis software package, allows this to be done. Figure 5-5 illustrates this, with speed traces from two different laps by Martyn Bell shown on the one graph. At the top you can see the lap times recorded (by the logger, not the time keepers) for each of the two laps (lap 4 was actually Martyn's best in this qualifying session). Straight away you can begin to see the differences in the speed traces in different parts of the course between the two laps. The red trace is the faster lap, and you can see that this lap was faster virtually everywhere. We'll go into this in more detail later.

Things start to get a little more complex as you overlay more channels, and Figure 5-6 shows two channels, speed and rpm, with the same two laps overlaid. The rpm trace looks interesting from around 1,250m into the lap to about the 2,000m point, but again we'll look into the detail more closely later. Some of the differences between laps obviously show up very clearly, although there are more subtle but nevertheless important differences to be found in the finer detail.

Figure 5-7 shows four channels plotted, with two laps overlaid. This is getting a bit busy now, and clearly the zoom functions, and the numerical values of each channel at the cursor displayed in the boxes on the right of the screen, become more useful in deciphering the detail. The Astratech software is again like many others in that it allows you to select which channels you want to display. This can make things much clearer, as Figure 5-8 shows. Different systems allow different degrees of flexibility in how you display your selected channels, but it's worth getting to know your chosen system in this respect so that you can make things as clear as possible for yourself. Here, just speed and throttle position have been plotted, with the same two laps overlaid. The zoom feature

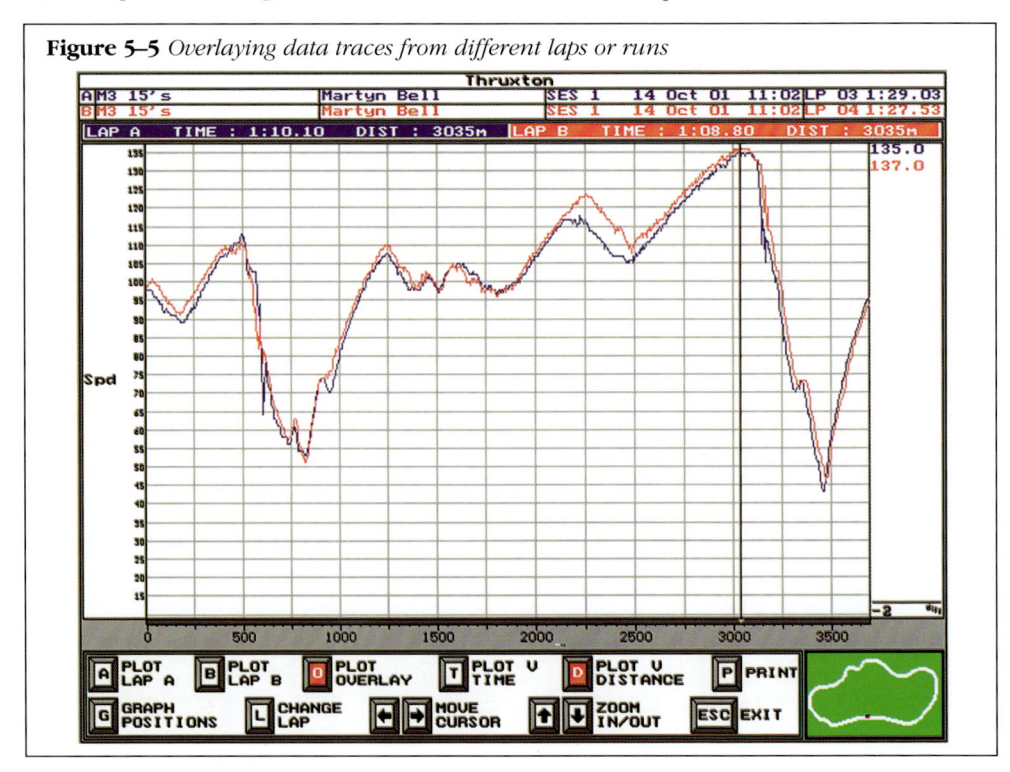

Figure 5–5 *Overlaying data traces from different laps or runs*

MORE CHANNELS! 61

Figure 5–6 *Overlaying the data from two channels*

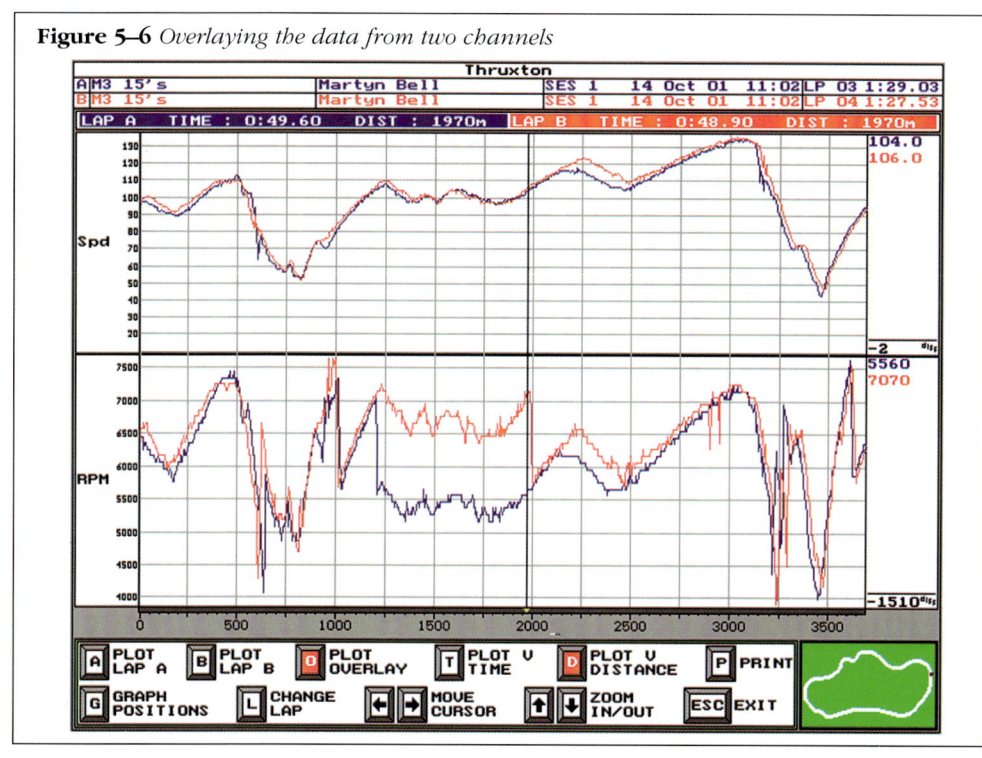

Figure 5–7 *Overlaying the data from four channels*

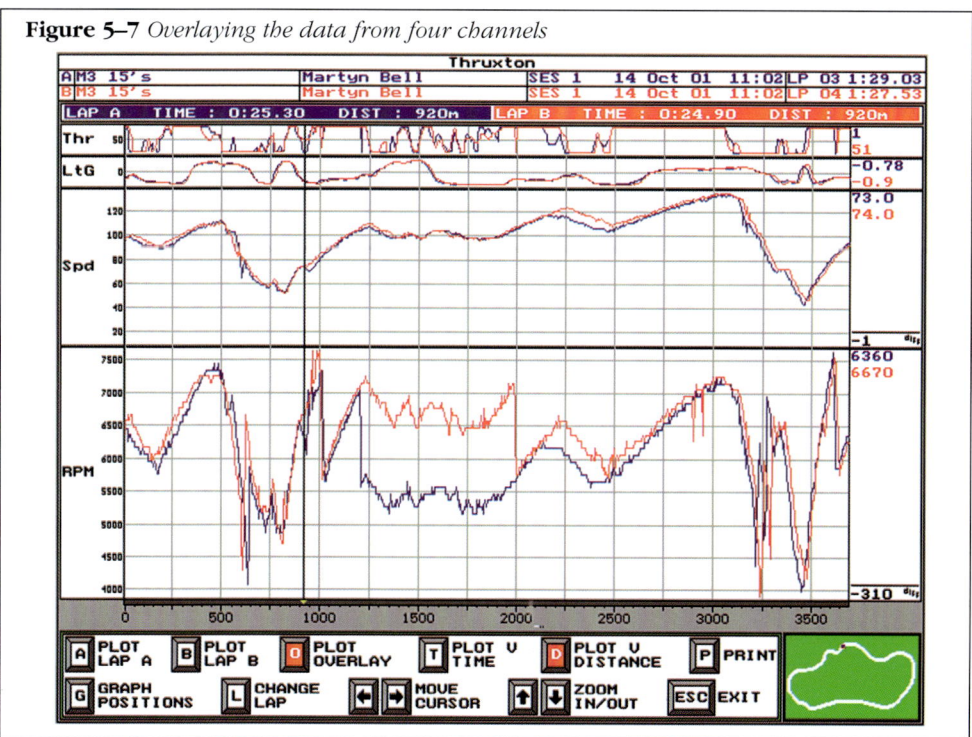

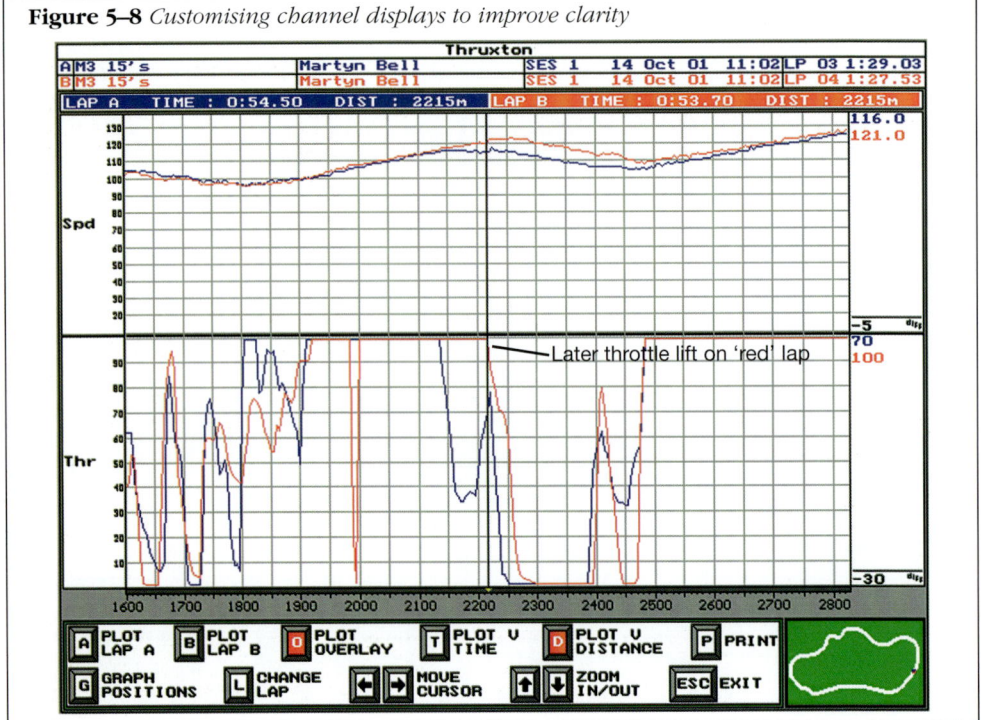

Figure 5–8 *Customising channel displays to improve clarity*

has also been used to enable the area mentioned above, the approach to Church Corner, to be examined more closely. It is clear, for example, that Martyn did stay on the throttle longer on the 'red' lap (for another 80 metres in fact), and this resulted in the 5mph advantage shown at the cursor. The speed advantage grew even bigger slightly deeper into this section.

Filtering, or data smoothing

Your DAS software allows you to 'smooth out' sometimes spikey graph lines with a 'filter' option. This invokes a mathematical averaging of the data and plots a smoother graph line, which can sometimes be useful. However, take care not to filter out real, high frequency data that might be trying to tell you something, and don't use filtering to make up for a poorly installed or faulty sensor. For example, placing your G sensors in the middle of an unsupported floor panel that vibrates will see the sensor logging those vibrations. Move the sensor rather than filtering the data. In general, use as little filtering as you can get away with.

More than just graphs

Graphic display of data is not just restricted to line graphs. Data analysis software can present logged information to you in a number of useful and informative ways, and it helps to be familiar with the different methods and how they interplay with the interpretation of the line graphs. Not all of these information displays are graphical, some involve tables of data but presented in a way that quickly summarises what has been going on. Again, the Astratech package will illustrate here, but other DAS software packages will do similar things.

Figure 5-9 shows one of the simplest reports, listing the lap times from a particular session. These can be displayed in run order or lap time order, with the fastest at the top of the list. The min-max report shown in Figure 5-10 is just that,

MORE CHANNELS!

Figure 5-9 *The simple lap time report*

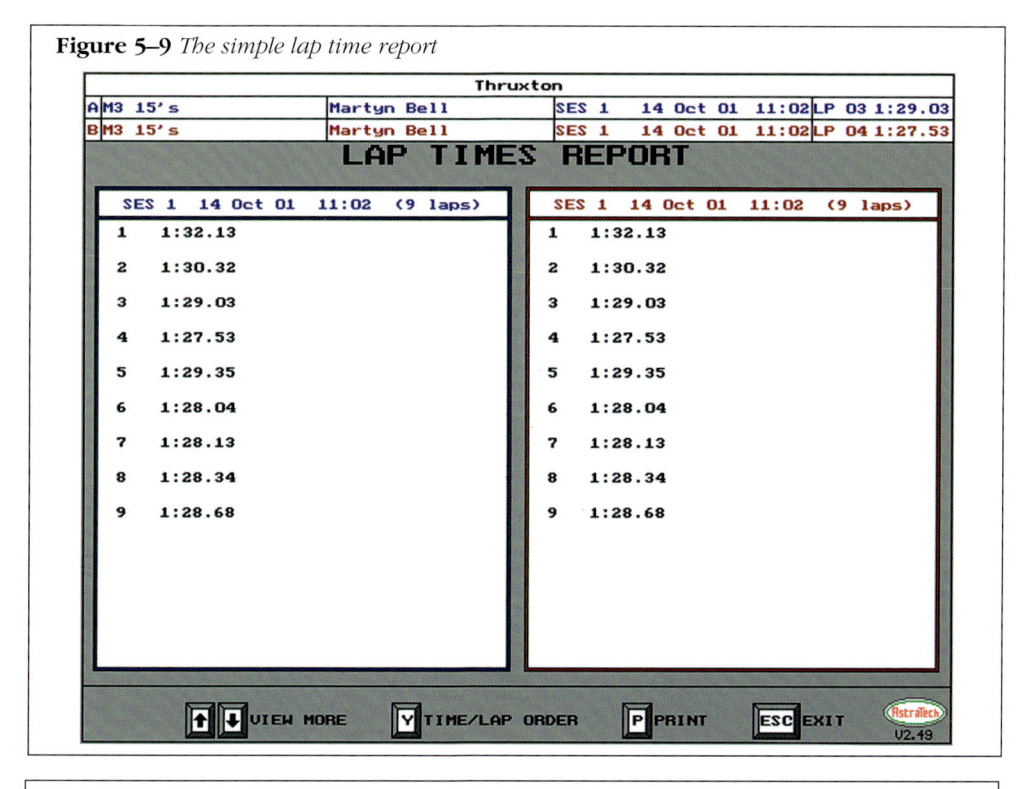

					Thruxton			
A	M3 15's		Martyn Bell		SES 1	14 Oct 01	11:02	LP 03 1:29.03
B	M3 15's		Martyn Bell		SES 1	14 Oct 01	11:02	LP 04 1:27.53

LAP TIMES REPORT

SES 1 14 Oct 01 11:02 (9 laps)	SES 1 14 Oct 01 11:02 (9 laps)
1 1:32.13	1 1:32.13
2 1:30.32	2 1:30.32
3 1:29.03	3 1:29.03
4 1:27.53	4 1:27.53
5 1:29.35	5 1:29.35
6 1:28.04	6 1:28.04
7 1:28.13	7 1:28.13
8 1:28.34	8 1:28.34
9 1:28.68	9 1:28.68

↑↓ VIEW MORE   Y TIME/LAP ORDER   P PRINT   ESC EXIT   AstraTech V2.49

Figure 5-10 *The Min-Max report*

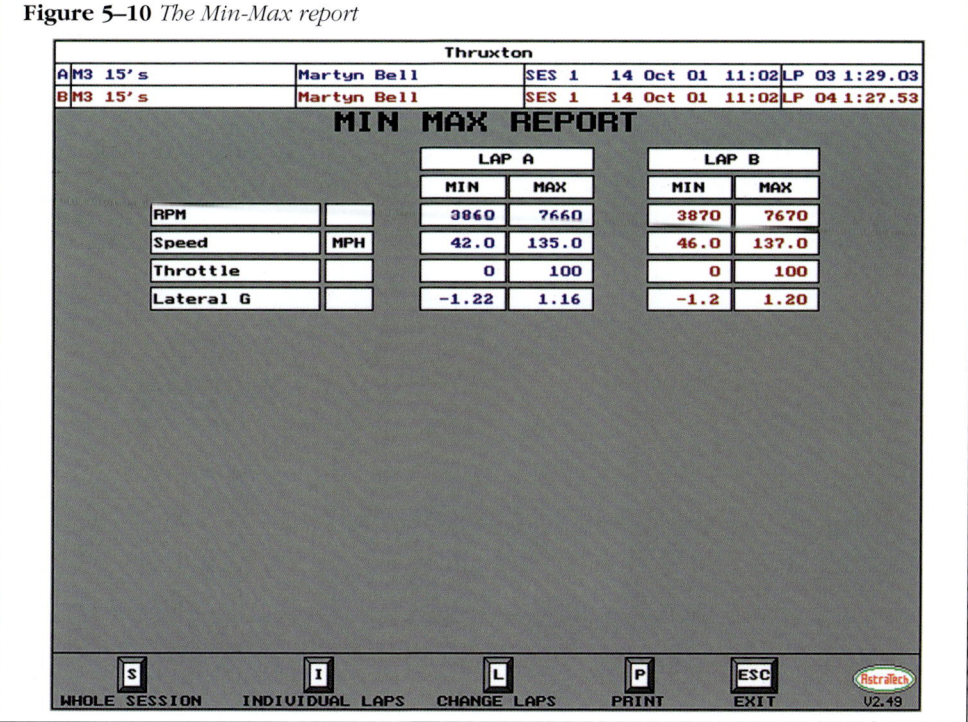

					Thruxton			
A	M3 15's		Martyn Bell		SES 1	14 Oct 01	11:02	LP 03 1:29.03
B	M3 15's		Martyn Bell		SES 1	14 Oct 01	11:02	LP 04 1:27.53

MIN MAX REPORT

		LAP A		LAP B	
		MIN	MAX	MIN	MAX
RPM		3860	7660	3870	7670
Speed	MPH	42.0	135.0	46.0	137.0
Throttle		0	100	0	100
Lateral G		-1.22	1.16	-1.2	1.20

S WHOLE SESSION   I INDIVIDUAL LAPS   L CHANGE LAPS   P PRINT   ESC EXIT   AstraTech V2.49

showing the highest and lowest values recorded on each channel. This might be useful when unusually low or high values of a parameter are flagged up, especially if the channels related to engine parameters for instance.

Figure 5-11 is an especially useful means of comparing times in segments of the track. With the Astratech software these segments are manually entered, and can be adjusted to what provides meaningful data. Thus, the segments I applied to Thruxton in these examples here relate to parts of the track any driver might be interested in – in essence the corners and the straights. It is possible to break the circuit down into more segments if required, but six to eight meaningful chunks is probably sufficient. The split times report shown in Figure 5-11 shows the recorded times in each of the track segments for each of the nine laps logged in this session. Some of the segments are off-screen to the right and the keyboard's cursor keys (the four arrows pointing up, down, left and right) are used to scroll to them, which is why they are not all visible in this screen shot. With this report you can see how the segment times compare lap to lap. Of most interest on this report are the lines near the bottom. The software adds together your best segments to compute the fastest lap you could do, if all your quickest segments were on the same lap! You can see here that Martyn's best actual lap was over half a second slower than his cumulative best segments time. The fastest rolling lap adds together the fastest consecutive group of segments, which here was from segment four of lap seven through to segment three of lap eight. This was just one hundredth faster than Martyn's best actual lap (lap four).

The segment report, shown in Figure 5-12, is a way of comparing the times and speeds in the track segments between selected laps. On the left of the table are the segments as defined through the software with reference to the graphs and a track map. The columns headed 'time

Figure 5–11 *The split times report*

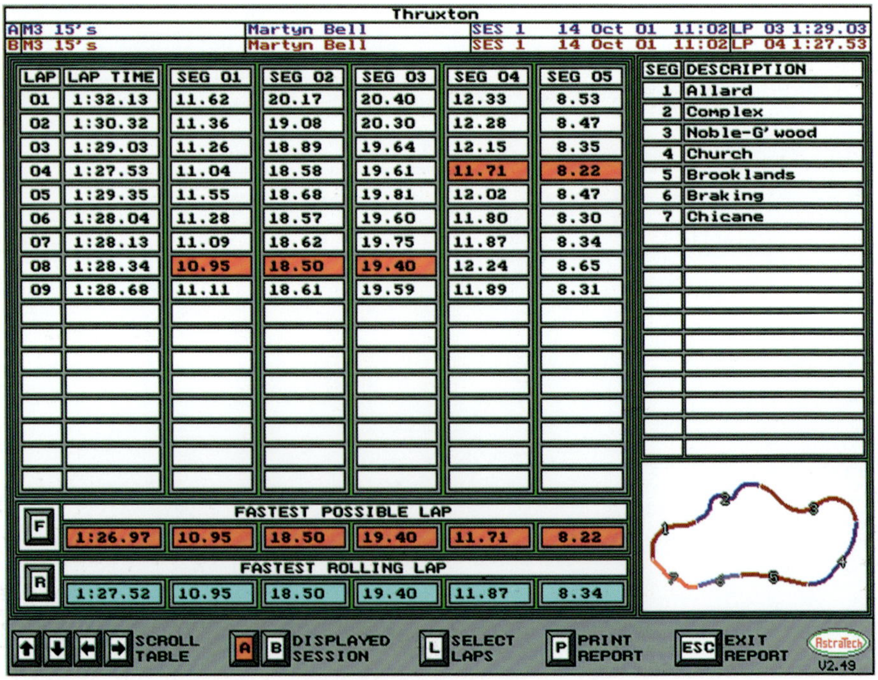

LAP	LAP TIME	SEG 01	SEG 02	SEG 03	SEG 04	SEG 05
01	1:32.13	11.62	20.17	20.40	12.33	8.53
02	1:30.32	11.36	19.08	20.30	12.28	8.47
03	1:29.03	11.26	18.89	19.64	12.15	8.35
04	1:27.53	11.04	18.58	19.61	11.71	8.22
05	1:29.35	11.55	18.68	19.81	12.02	8.47
06	1:28.04	11.28	18.57	19.60	11.80	8.30
07	1:28.13	11.09	18.62	19.75	11.87	8.34
08	1:28.34	10.95	18.50	19.40	12.24	8.65
09	1:28.68	11.11	18.61	19.59	11.89	8.31

SEG	DESCRIPTION
1	Allard
2	Complex
3	Noble-G'wood
4	Church
5	Brooklands
6	Braking
7	Chicane

FASTEST POSSIBLE LAP
F | 1:26.97 | 10.95 | 18.50 | 19.40 | 11.71 | 8.22

FASTEST ROLLING LAP
R | 1:27.52 | 10.95 | 18.50 | 19.40 | 11.87 | 8.34

Figure 5–12 *The segment report*

		Thruxton				
A	M3 15's	Martyn Bell	SES 1	14 Oct 01 11:02	LP 04 1:27.53	
B	M3 15's	Martyn Bell	SES 1	14 Oct 01 11:02	LP 08 1:28.34	

SEGMENT REPORT

NO	SEGMENT DESCRIPTION	POSITION START	POSITION END	TIME GAINED BY A	TIME GAINED BY B	ENTRY SPEED A	ENTRY SPEED B	EXIT SPEED A	EXIT SPEED B	MIN SPEED A	MIN SPEED B
1	Allard	0	490		0.09	97.5	95.6	109.6	109.4	91.9	95.6
2	Complex	490	1095		0.08	109.6	109.4	96.4	95.0	51.8	55.5
3	Noble-G'wood	1095	1985		0.21	96.4	95.0	107.2	103.6	96.4	94.6
4	Church	1985	2590	0.53		107.2	103.6	115.1	106.2	107.2	102.0
5	Brooklands	2590	3055	0.43		115.1	106.2	135.9	133.0	115.1	106.2
6	Braking	3055	3330		0.14	135.9	133.0	72.9	77.1	72.9	77.1
7	Chicane	3330	3715	0.37		72.9	77.1	96.0	94.9	47.0	42.0

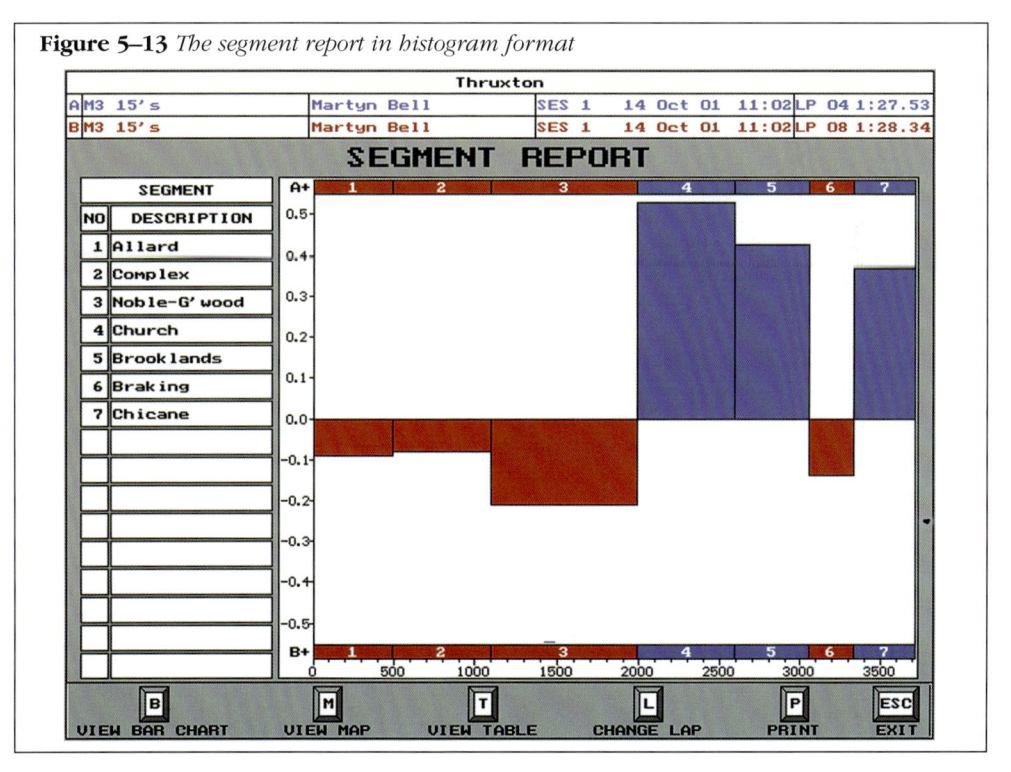

[B] VIEW BAR CHART [M] VIEW MAP [T] VIEW TABLE [L] CHANGE LAP [P] PRINT [ESC] EXIT

Figure 5–13 *The segment report in histogram format*

		Thruxton				
A	M3 15's	Martyn Bell	SES 1	14 Oct 01 11:02	LP 04 1:27.53	
B	M3 15's	Martyn Bell	SES 1	14 Oct 01 11:02	LP 08 1:28.34	

SEGMENT REPORT

NO	DESCRIPTION
1	Allard
2	Complex
3	Noble-G'wood
4	Church
5	Brooklands
6	Braking
7	Chicane

[B] VIEW BAR CHART [M] VIEW MAP [T] VIEW TABLE [L] CHANGE LAP [P] PRINT [ESC] EXIT

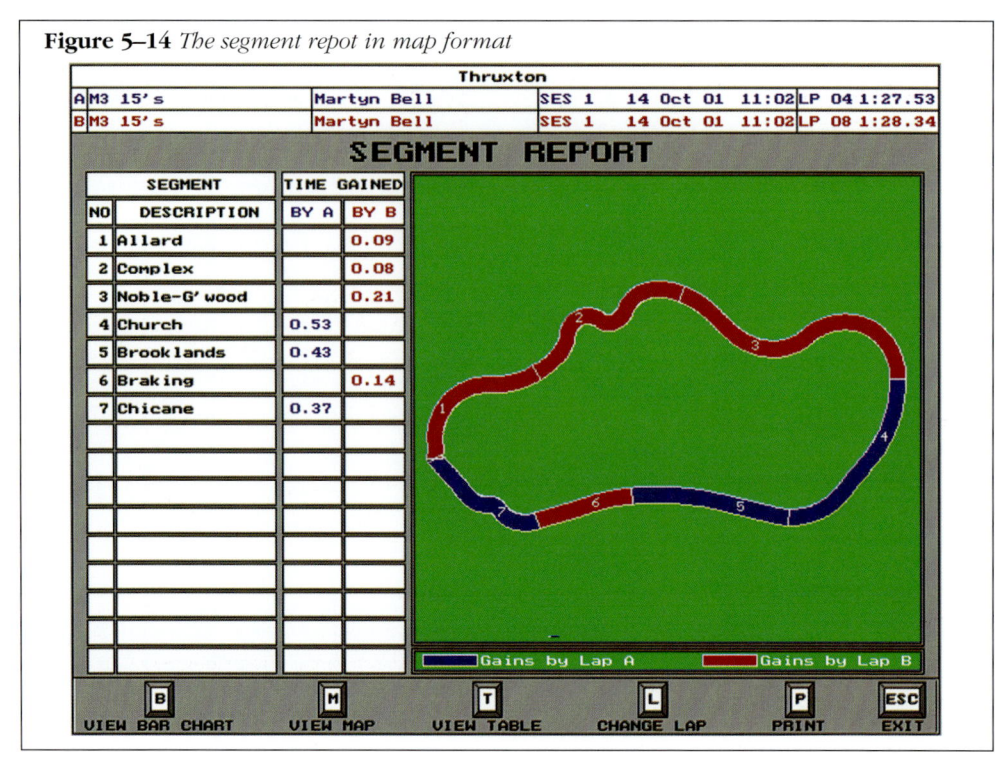

Figure 5–14 *The segment repot in map format*

		Thruxton			
A	M3 15's	Martyn Bell	SES 1	14 Oct 01 11:02	LP 04 1:27.53
B	M3 15's	Martyn Bell	SES 1	14 Oct 01 11:02	LP 08 1:28.34

SEGMENT REPORT

SEGMENT		TIME GAINED	
NO	DESCRIPTION	BY A	BY B
1	Allard		0.09
2	Complex		0.08
3	Noble-G'wood		0.21
4	Church	0.53	
5	Brooklands	0.43	
6	Braking		0.14
7	Chicane	0.37	

Gains by Lap A | Gains by Lap B

B	M	T	L	P	ESC
VIEW BAR CHART	VIEW MAP	VIEW TABLE	CHANGE LAP	PRINT	EXIT

gained by A or B' show where and how much time was gained on one lap compared to another. The columns allowing comparisons of entry, exit and minimum speeds are especially valuable when looking closely at corner approach, speed through the apex, and corner exit. This could have value on driver, chassis, engine and aerodynamic performance analysis, to name but a few areas of interest!

The segment reports can also be presented in different ways. Figure 5-13 is a segment report in bar chart format, which immediately points to the segments where most time was gained or lost. Figure 5-14 is the same information presented as a track map, which shows which segments on the track were faster on a particular lap.

Other graphical information that the software can provide includes histograms of the logged channels, as discussed in Chapter 3. But finally, one of the graphical displays that Astratech can claim to have pioneered is the use of animated icons to demonstrate what was happening to each channel. Devised because they are easier to interpret than graph lines, these animated icons, known by Astratech under the trade name of Track-a-mation, can be helpful, at least when you first get involved with data logging and are getting used to interpreting the graphs – see Figure 5-15.

The symbols here represent a rev counter, a digital speedometer, a foot on a throttle and a swinging pendulum for lateral G. On the track map (bottom right of the track) are two small circles, representing the car on lap A and lap B, and it is possible to use the software to 'drive' the cars around simultaneously a step at a time using the cursor, or by using a 'free run' facility, which plays back the displayed laps in a kind of fast-forward mode. The data is played back in increments of a tenth of a second (the Astralog DAS records at 10Hz, or 0.1 second intervals) so that the relative positions on the track versus time may be seen, along with

Figure 5–15 *'Track-a-mation'*

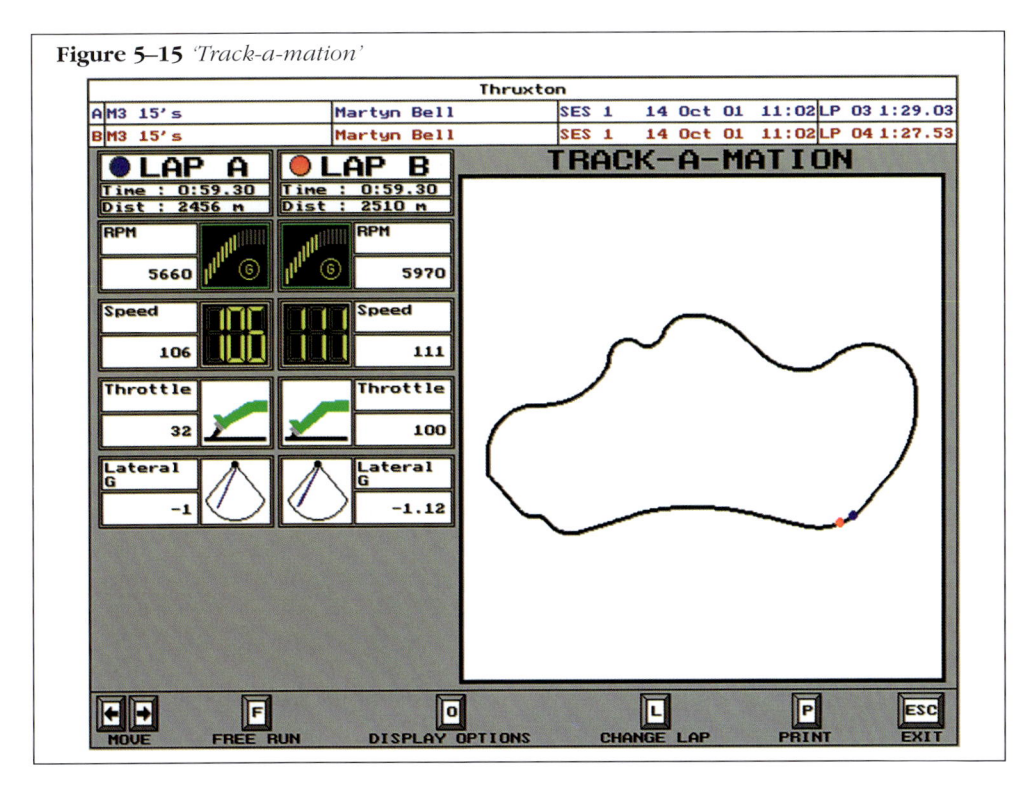

what was happening on each recorded channel. It becomes very evident where the car on one lap pulled ahead compared to the other lap, and it is also easy to see, more in a qualitative than a quantitative way, even though the channel values are displayed, what was happening on each channel. They are a quick way of helping to focus on some of the major events that each channel records.

Other more expensive and sophisticated DAS software packages will offer yet more ways of displaying information. I'd hazard a guess though that once you're familiar with your DAS that you'll probably single out just a very few ways of examining the data that suit you best.

Let's now get into some more detailed data interpretation using real examples. We'll take it a sensor at a time, but as we go along it will be apparent that what is happening on each channel is very often related to what is happening on the others. This is not surprising when you think about it – a competition car/driver system is made up of what might be termed sub/systems that interact with each other. The driver presses the throttle, which accelerates the engine, which in its turn governs what the chassis does, for example, and it is by thinking about the interrelationships of the channels you are logging that their interpretation becomes easier. We'll begin with what it's all about – speed.

Chapter 6

Logging speed

AS ALREADY STATED, speed is your most useful and important channel. It tells you what you really want to know at every point of the track – in other words, how fast are you? Logged speed, in fact, relates to every aspect of car performance: acceleration, braking, chassis performance and set-up, engine performance, aerodynamic performance and driver performance. All these aspects are inter-related, but study of the speed trace can point you in a direction of enquiry when analysing your overall performance. Comparison between laps or runs is what brings the most value though, whether it's comparing with your own earlier performance, or with someone else's. You can see the effects of changes made, whether to the car, the engine or the driver. Lap timing summarises the whole lap, but looking at particular sections of a speed trace (or any data chan-

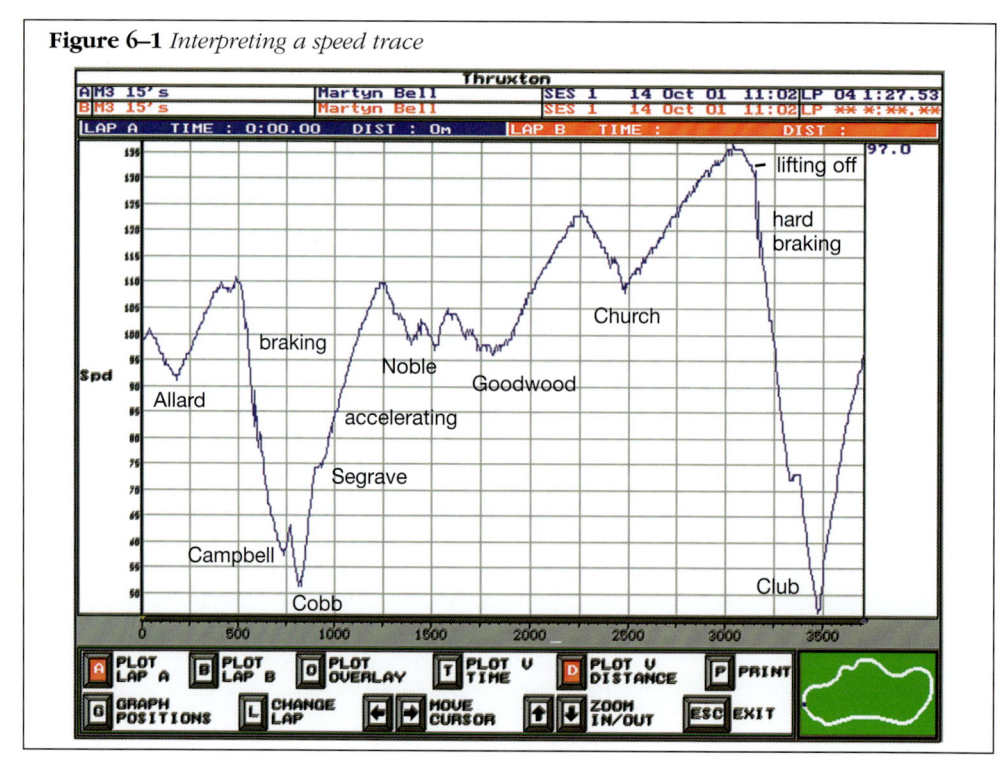

Figure 6–1 *Interpreting a speed trace*

nel trace come to that) can tell you how you and your car performed in, say, one fast corner after a modification or a set-up change was made.

Speed data interpretation

Let's do a lap of Thruxton with Martyn Bell in his BMW M3 again, and go through each part of the track to see what's happening on the speed trace. Figure 6-1 once more shows a single speed trace, annotated to point out the main geographical locations and what the car was doing at those points.

The start of the lap is at the extreme left-hand side of the graph, and as discussed previously the graph baseline is in units of distance (metres). Just after the startline, or rather his lap beacon, Martyn peaked at 101mph before slowing for Allard Bend, which like some other of Thruxton's corners, is a fast right-hander. The downward slope of the trace showing the car slowing is indicative of either gentle braking or simply lifting off the throttle for this corner. The first speed trough, which corresponds to the slowest speed in Allard (probably at or just before the actual corner apex) shows a minimum speed of 91mph in this corner.

The upward slope of the trace then shows the car accelerating out of Allard, around what is actually a sweeping, gentle left-hand curve. The speed peaks at 111mph, but it looks as though Martyn slowed and then gained speed again just before the maximum speed was reached here – we'll come back to that later. After this peak, the speed drops rapidly on the approach to the Complex, an indication of braking. At this point it is possible to use the calculation outlined in Chapter 4 to work out the longitudinal G-force generated under the brakes. At the steepest part of the speed trace the car slows from 109mph to 87mph in 1.8 seconds, corresponding to -0.56G deceleration. If this doesn't sound like rapid slowing down, remember that the course is still turning here.

The Complex, as the track map shows, is a right–left–right sequence, and the car dips to 57mph in Campbell, the first right-hander, then 51mph in Cobb, but then accelerates rapidly towards Segrave for which the speed merely levels off at 74mph briefly before climbing again. The slope of the speed trace after Cobb is quite steep and shows a gain of 22mph in 2.6sec, equal to 0.39G.

After Seagrave the speed trace continues climbing, but at a slower rate (as aerodynamic drag builds) as the car heads towards Noble, a fast double-apex left-hander. There is a short dip in the speed trace at 985 metres, hard to see at this scale but more obvious when zoomed in (see Figure 6-2), which actually corresponds to an up-shift to fourth gear. The rate of acceleration in the higher gear is also less, and decays further (the slope gets less steep) as speed increases. The speed then climbs to 110mph before Martyn slows somewhat for the first part of Noble. Again, the downward slope indicates lifting off rather than braking. We'll come back to this part of the trace later too.

The speed dips to 98mph in the first apex of Noble, climbs briefly, then dips to 97mph before rising fairly rapidly though briefly to a peak of 105mph before slowing for Goodwood. This is another fast right-hander taken here at 96mph before the car then accelerates for quite a long time around Village Curve, peaking at 124mph. Once more the car slows as Martyn prepares himself and the car for the infamous Church Corner, a very quick right-hander with a nasty bump at the apex. The preparation consists of a longish (and perfectly understandable, if you've driven the track . . .) throttle-lift by the look of the speed trace. The speed dips to 108mph in Church.

The long drag from Church, through Brooklands and up Woodham Hill shows a more or less linear increase in speed (interesting in itself), and the maximum speed for the lap (137mph) is reached prior to slowing for the tight Club Chicane. You can see from the way the

Figure 6–2 *Detail shown when zoomed in*

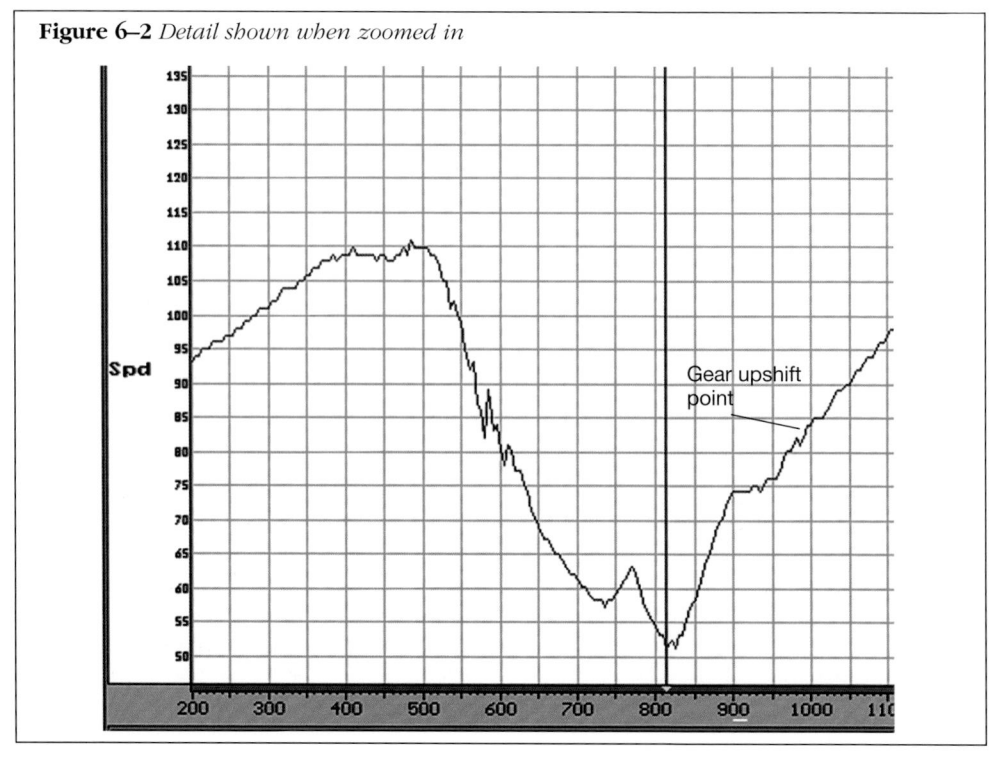

[Speed trace chart with Y-axis labeled "Spd" ranging from 50 to 135, X-axis from 200 to 1100. Annotation "Gear upshift point" pointing to a feature in the trace.]

trace slopes gently at first and then very steeply that Martyn must have lifted off the throttle at first but didn't get on the brakes immediately. However, when he did brake the speed decreased pretty rapidly (–0.77G for the steepest part of the speed trace).

The speed drops down to a circuit-minimum of just 46mph at the apex of the slowest part of the Club Chicane. However, there is what at first is an odd-looking 'bump' in the slow-down at 3,330 to 3,380 metres where the speed appears to briefly climb and level slightly before finally dipping to the minimum. This is actually where the car is being driven around the first, right-handed part of the Club Chicane. The speed is then reduced quite a bit further, by braking going by the downward slope of the speed trace, towards the tight left-handed part of the chicane.

Once through the slowest part of the Club Chicane (which is the left-handed part) the speed trace increases rapidly again as the car accelerates hard out of the chicane to head past the lap beacon again.

So that's a complete lap of Thruxton in a championship-winning BMW M3! And it was the quickest lap of the session for Martyn, which actually gave him pole position for the race later that day. But from our point of view, it helps to go right through a trace in this way, especially using the speed channel, in order to fix the peaks and troughs in your mind as to where on the course you are. Yes, there's a little dot moving on a map of the track to further help with this, but by getting familiar with the characteristic speed trace shape for the venues you visit, which you will over time, you will be able to analyse and compare laps or runs much more quickly. We'll prove that point now by looking at the benefits of overlaying data from different laps. Figure 6-3 shows two of Martyn's laps overlaid to provide comparison. Without the benefit of the annotations used previously, it will now be possible to relate the peaks and troughs to the course with relative ease . . . But feel free to flip the pages and refer back to Figure 6-1 to make sure!

Data overlays

This is where the real value of data logging makes itself apparent. Looking at a single trace in isolation can be very informative. For example, if you've never been to a track before and somebody is good enough to provide you with a speed trace from their car, if you know how your car performs relative to theirs you can get a very good idea of the speeds you'll be doing. This will help you pick a set-up for the car for your first visit – and in Chapter 13 we'll look at a really interesting example of how the Champ Car teams used just this approach when they first travelled to Europe to race at Lausitz in Germany and Rockingham in England in 2001.

But when it comes to analysing how you can improve, or how you have improved the performance of you and your car, then data comparison is invaluable. So, let's take a look at how you make these comparisons, and what you can glean from them. Again, to keep it simple we're only looking at speed for now. In Figure 6-3 are two consecutive laps by Martyn Bell. The blue trace shows his fastest qualifying lap, which we've already looked at, and the red trace is the previous lap, his third full lap out of the pits. Just let your eye follow the traces for a few seconds so you can pick out some of the obvious similarities and differences, and then come back for a section-by-section tour of the two traces.

OK, the first thing you might have noticed is the basically similar ups and downs, synchronised by the fact that the corners are all in the same place on the track, and the lines Martyn took around the two laps were fundamentally the same. Then, what is very apparent is that in places the traces are quite different, for example from the 2,250 to the 2,500-

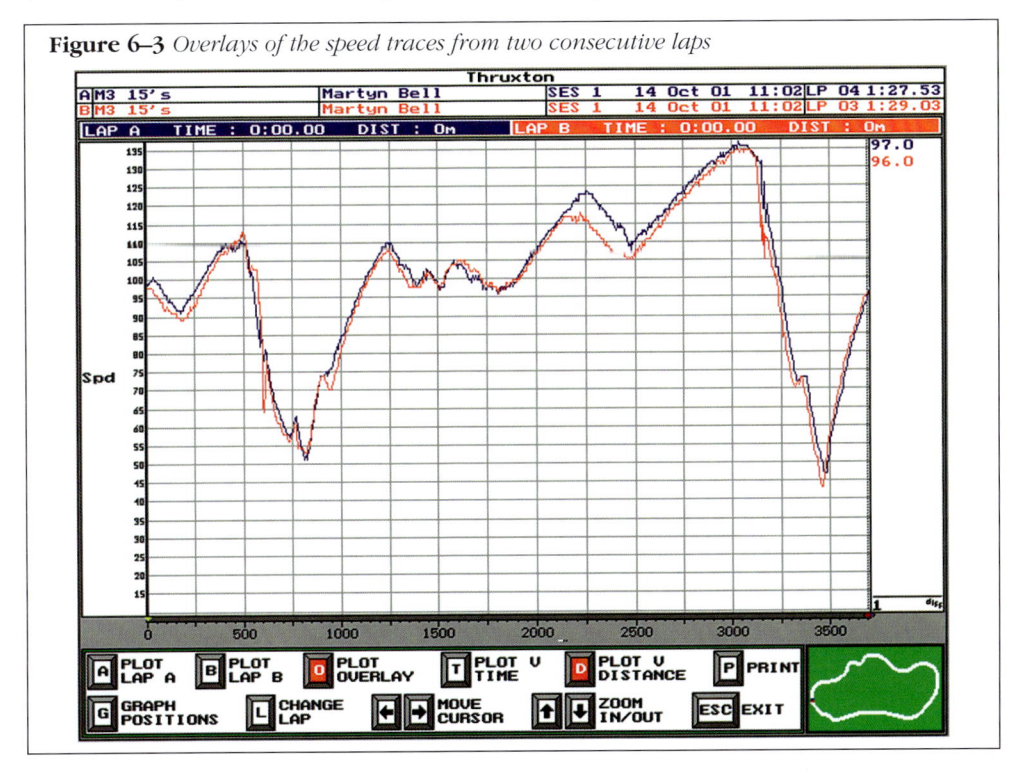

Figure 6–3 *Overlays of the speed traces from two consecutive laps*

metre points. So let's start at the beginning and work our way around the two laps to examine the similarities and differences in more detail.

Remember that these were consecutive laps, the 'red' lap leading into the 'blue' lap. So, right from the start the car was travelling faster on the blue lap. This actually goes back to the end of the red lap, where Martyn has driven through the chicane very slightly slower (by 3mph in fact), but has also clearly started accelerating a little sooner too. The difference is around 10 or 15 metres, and suggests the slower run through the chicane enabled the car to hold a tighter, cleaner line and let Martyn get on the gas that bit sooner. This allows him to peak 3mph faster as he passes his lap beacon. But in fact he carries that extra speed into, through and out of Allard. So the blue lap should have been faster in that first segment. You could check this out with the segment reports.

The blue trace shows the extra speed being carried into the next segment as the car continues to gain speed, but then as it peaks prior to braking for the complex you can see that the approach to the braking point was slightly different, possibly slightly more conservative, on the blue lap. In fact the red lap peaks at 113mph here compared to 110mph on the blue lap, but the car begins to slow down at exactly the same point. Now, look carefully at the red trace. It shows a couple of short, near vertical drops in speed, after which the trace seems to 'recover' to the normal down-slope. This is caused by the wheel on which the speed sensor was fitted, locking up briefly under braking. In other words, on the red lap Martyn was on the limit of adhesion as he braked for the Complex. The blue trace here shows a smoother loss of speed, and in fact ends up with a little more speed being carried into the apex at Campbell. This is a classic case of the 'last of the late brakers' (Martyn on his red lap) being out-gunned by a smoother, less frenetic braker (Martyn on his blue lap). So, you can see from this comparison that a little time was gained here on the blue lap.

The speeds out of Campbell and into Cobb were pretty similar, but look at the difference at Segrave. On the red lap Martyn obviously slowed for Segrave, dropping about 4mph as he headed towards the apex. But on the blue lap, although the car stopped gaining speed, it didn't lose any speed, and the result of that was that the car carried more speed through the corner, and from the corner exit and along the next 'straight'. Overall then the passage through the Complex segment should have been quicker on the blue lap.

The next section, Noble to Goodwood, is interesting and appears to show gains and losses on each lap. Prior to slowing for, and into the first part of Noble the blue lap is slightly faster. But by the time the car on each lap has reached the first apex of Noble the speeds were very similar, and then on the way into Goodwood the car carried a little more speed on the red lap. Exiting Goodwood saw the laps gain speed at a very similar rate. So overall through this segment the average speed was probably pretty similar.

The next segment, which includes the run into and through Church Corner shows a big difference between the laps. Quite obviously Martyn has kept the speed increasing for longer on the blue lap, with the result that he peaked at 124mph at the 2,260m point on the blue lap compared to 118mph at the 2,225m point on the red lap. An 8 to 9mph advantage was carried to the apex at Church, and a small advantage in apex speed (2 to 4mph) was also maintained, leading to a faster exit speed. So the blue lap ought to have produced a significant gain in time compared with the red lap.

The extra exit speed from Church can be seen to have stayed with the car on its run all the way through Brooklands and up Woodham Hill to the braking area for the Club Chicane. An extra 3–4mph was carried pretty well the whole distance, with peak speed being 2mph greater on the blue lap. This represents a small percentage gain relative to the high speeds being achieved here, but is nevertheless a gain for the blue lap, and demonstrates the importance of exit speed from a corner on the speed at the end of the ensuing straight. All too often drivers who achieve poor straightline speed will blame either the engine (down on power) or the aerodynamics (too much drag) when the cause is actually their exit speed off the previous corner. Of course, this might be related to poor chassis or aerodynamic efficiency in that corner instead of driver trouble!

The final segments of the lap include the braking area for the Club Chicane and the chicane itself. On both laps the driver appears to have started slowing at pretty much the same point on the circuit, but then maybe braked a little earlier and a little harder on the red lap, this trace being slightly slower all the way to the slowest point of the Chicane. And this is where we came in, because the speed picked up more rapidly from the Chicane on the red lap, which led to that early gain in speed on the blue lap. Did he do as well as he could have? Well, remember Figure 5-11 showed the best theoretical lap time, calculated from the best segment times actually set by Martyn in this session? It was over a half second quicker than the blue lap shown here, which was the best overall lap in reality. It just goes to show, you can always improve!

It was worth taking the time to run around these two laps to see how informative comparing speed traces is. But clearly these comparisons also raise as many questions, and they supply answers. Where differences in speed show up on the trace, the temptation is

Driver feedback is vital to good data interpretation.

to try to supply a speculative answer straight away. This might be the right answer, or it could be way off. The point where Martyn's blue lap was so much faster approaching Church Corner could very well be because he kept his foot on the throttle for longer. But that might have been because he was looking in his mirror or at his instruments and missed his slow-down point! So don't always be tempted to rush to find the answers – just make a note of the questions intially; 'Why was the driver so much quicker here, or slower here?' By asking the questions in driver debriefs, or using the data to remind yourself if you are a one-person team, you'll learn a lot more than by supplying rapid conjectural and possibly completely erroneous answers.

Of course, there are many possible reasons why one lap may show differences in speeds relative to another lap. In the example we looked at here, the laps were consecutive and contiguous, and therefore it would not have been possible to make many chassis set-up changes between them. There might have been the possibility for the driver to alter anti-roll bar stiffnesses to alter the handling balance, or to adjust the front to rear brake balance, but certainly no aerodynamic changes could have been made. Similarly, engine modifications could at best have only been minor, had there been the option for the driver to adjust or change the engine management system map from the cockpit, for example. So we can conclude fairly safely in this case that the differences in speeds were made by the driver. That means the questions we ask will be related to what the driver did. However, if the two laps came from different sessions at a test, and for example the rear wing had been set to a slightly higher angle for the blue lap, then we might easily draw different conclusions from the comparisons. So you also need to know what the circumstances were in which the data was logged to be able to draw meaningful conclusions.

Combining speed and rpm

It's true that analysing a speed trace tells you a great deal of what you want to know about performance. But by cross-referring to other recorded channels, not only can you supply answers to some of the questions that arise, but you can also divine additional detail that raises further questions that lead, ultimately, to more answers and a better understanding of performance factors.

Figure 6-4 shows once again a lap of Thruxton with speed and rpm displayed. We've seen this graph before, but now we can examine it with more authority and obtain additional information, some of it basic, some of it of the 'what happened there?' variety. For the first part of the lap the rpm curve simply mirrors the speed curve as you'd expect it to here. The first detail of interest is the downshift that occurs at 615 metres during the slow-down for the Complex. As discussed in Chapter 4, the rpm trace can be used to see if the driver is blipping the throttle (heel and toeing) on the downshift, and there is no sign here, or at other downshifts, of an extra rpm peak associated with a throttle blip. We'll come back to this in the next chapter when we look at throttle position logging.

Moving on a bit further, there is an interesting portion to analyse starting at the 935m point. The rpm trace here follows its own path and is not mirrored by the speed trace for the next 30 to 40m. The car is seemingly at or just past the apex of Segrave Corner, and the car is accelerating from that point. The revs suddenly jump up, as though the load was rapidly reduced or removed from the engine. So what could have caused this trace? Could it be oversteer, with the rear end of the car swinging out and the associated lack of traction creating the brief over-rev? Or, since the rise in revs is very rapid and appears to coincide with the corner apex, could it be that the car ran over the kerb at the corner apex and momentarily lifted a rear (driven) wheel sufficiently to off-load the engine? From

Figure 6–4 *Interpreting speed and rpm together*

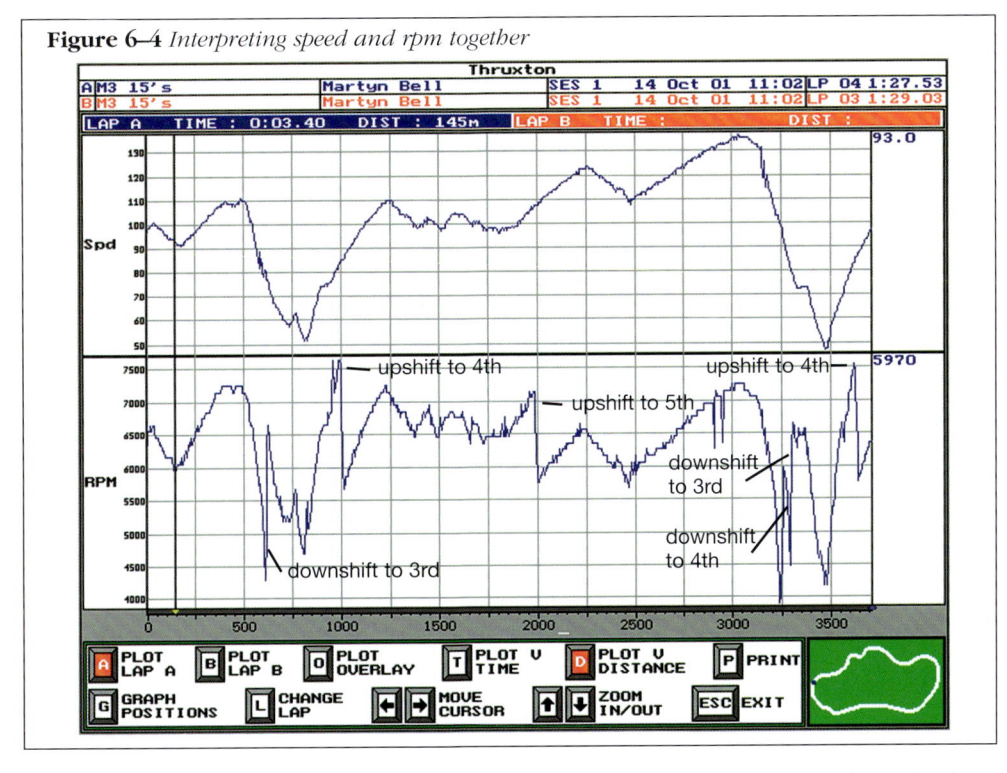

these two channels we can only spot the anomaly and ask the questions although without the ability to ask the driver at this point, my guess would be the latter because of the rapidity of the rev rise. But what is also apparent is that Martyn changed into fourth gear immediately after this little event, so he was certainly busy!

One way to attempt to narrow down the causes of apparently anomalous events is to see if they occur frequently or consistently. If this same pattern was repeated on every lap then it might increase confidence in the diagnosis. It would seem unlikely that any but an over-spirited, possibly even untidy driver would be hanging the tail out in exactly the same way on every lap in this corner, although there are some who enjoy themselves in this way. But running a kerb is something that is more likely to be repeated regularly as the driver seeks the quickest route around the track. In fact, a similar pattern was repeated on three out of nine laps in this session, which doesn't really help with diagnosis on this occasion! The only thing to do here would be to speak to the driver immediately after the data was examined (which would be immediately after the session of course . . .) and see what he recalls – hopefully something useful. In fact, months after the race meeting Martyn was able to recall kerbing the car occasionally.

Moving on to the 1,985m point, there's a detail that couldn't have been picked up from the speed trace of this lap alone, and that's the change from fourth to fifth gear. The rpm trace makes it quite clear. In itself this is not necessarily a crucial bit of information, but compare with lap 8 for example (Figure 6-5), and you can see a very different story at this point. Martyn has actually changed into fifth gear at the 1,225m point and driven Noble and Goodwood in top gear instead of fourth, as he did on lap 4. Why would he do this? I asked him, and he said it was to try to settle the car a little for this fast, 'knife-edge' curve, instead of running

Figure 6–5 *Two quick laps using different gears*

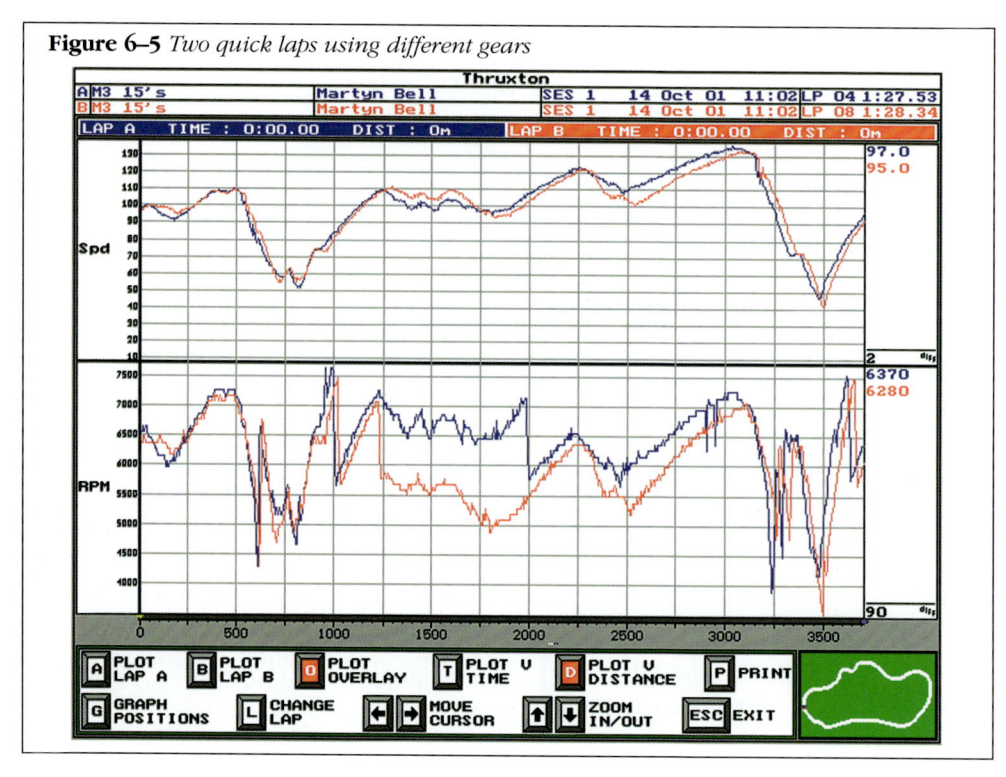

it at quite high rpm in fourth gear. From the speed trace you can see he actually went through Noble faster in the taller gear. The downside though was slower acceleration out of Goodwood towards Village. However, in the Noble-Goodwood segment using the higher gear gained 0.21sec according to the segment report, so the different technique was worth trying.

There are two downward spikes in the rpm trace at around 2,900 and 2,945m. These are probably just spurious signals rather then anything significant, although if they recurred then they could be indicative of some kind of gremlin in the electrics, either in the DAS or the ignition system.

Some of the occurrences discussed in this chapter become a whole lot clearer when the throttle trace is examined, which is what the next chapter is all about.

Chapter 7

On the throttle

FIGURE 7-1 SHOWS THE throttle position trace from Martyn Bell's fastest lap at Thruxton in the session we've been looking at. This is the same lap as the one we've been examining over the last few pages, but you'd be hard pushed to tell that by examining this graph! The fact is, looking at the throttle trace in isolation is not much help, yet using it in conjunction with other data traces can be highly revealing of both chassis and driver performance.

But first, just what does the throttle trace show? Look at the vertical, or y-axis on the graph. It goes from zero at the bottom to 100 at the top, although the axis markers only go from 5 to 95. Thus, throttle position is displayed in per cent, 100 per cent means the throttle is fully open, 0 per cent means it is fully closed, and obviously any value in between represents part throttle opening. This graph measures what your right foot is actually

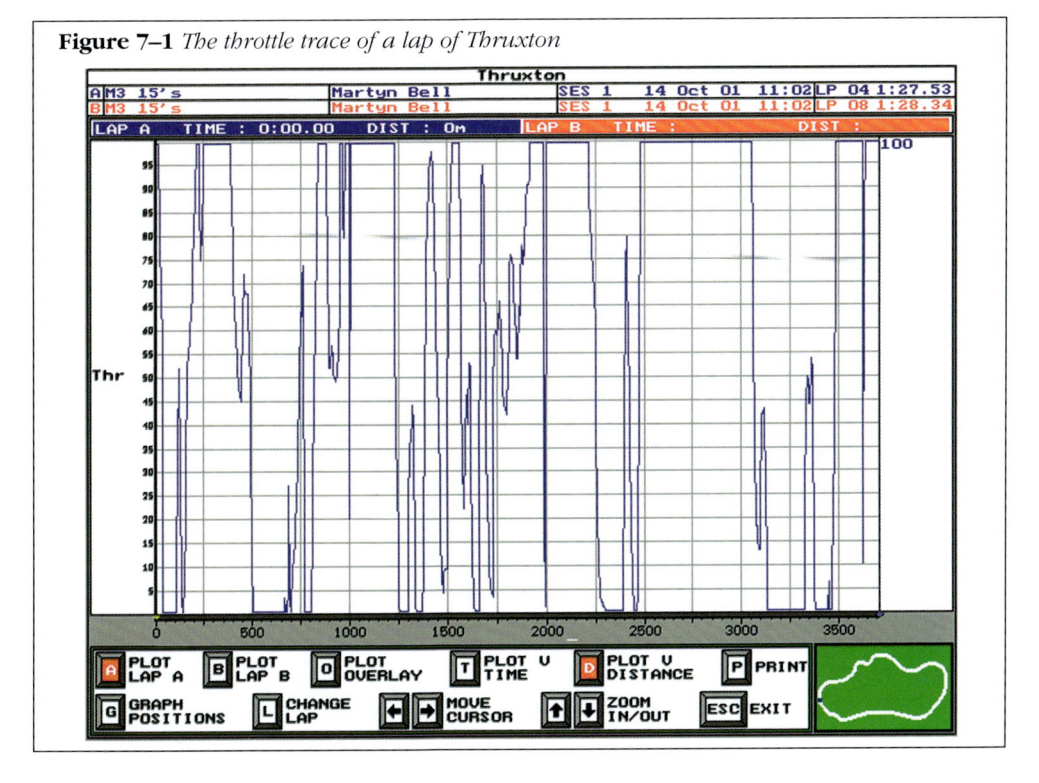

Figure 7–1 *The throttle trace of a lap of Thruxton*

doing on the loud pedal, as distinct (perhaps) from what you'd like to think it's doing! If you reckon you went through Buttock Clencher Curve with your pedal to the metal, this channel will either confirm it or not. But whatever you do, don't regard the throttle position sensor as the 'spy in the cab', because, like your DAS system itself, it may just turn out to be a best friend. You might not always like what it tells you, but you can rely on it to be honest with you! The fact is that in most situations you will do quicker times if the throttle spends longer in the fully open position. Getting on the throttle earlier in a corner can help and can hinder, coming off the throttle later going into a corner can do likewise, but generally speaking the more time the throttle is fully open, the more time you spend accelerating as hard as you can.

As with the other channels we've looked at, the throttle position trace is plotted out against distance (usually) or time on the horizontal, or x-axis. But to begin extracting any value from the throttle trace, we really do have to plot it with at least one other channel that enables us to 'navigate' where we are on the track. Figure 7-2 again shows the same 1min 27.53sec lap we've been studying up to now, but this time with rpm and throttle position plotted. Some of the main corners are marked again to help.

Running through a lap again, we'll quickly highlight some of the basic points in order to interpret what's going on with the throttle. The whys and wherefores can wait. Right at the start of the lap you can see that Martyn comes off the throttle (and on to the brake) as he slows for Allard Corner. The throttle trace drops to zero and the rpm fall in direct response. Just before the minimum rpm point in the corner though, it looks like Martyn tries to get on the throttle, thinks better of it, backs off again briefly and then is back on the throttle again, the revs now rising in response. This on–off–on occurrence might warrant further enquiry . . .

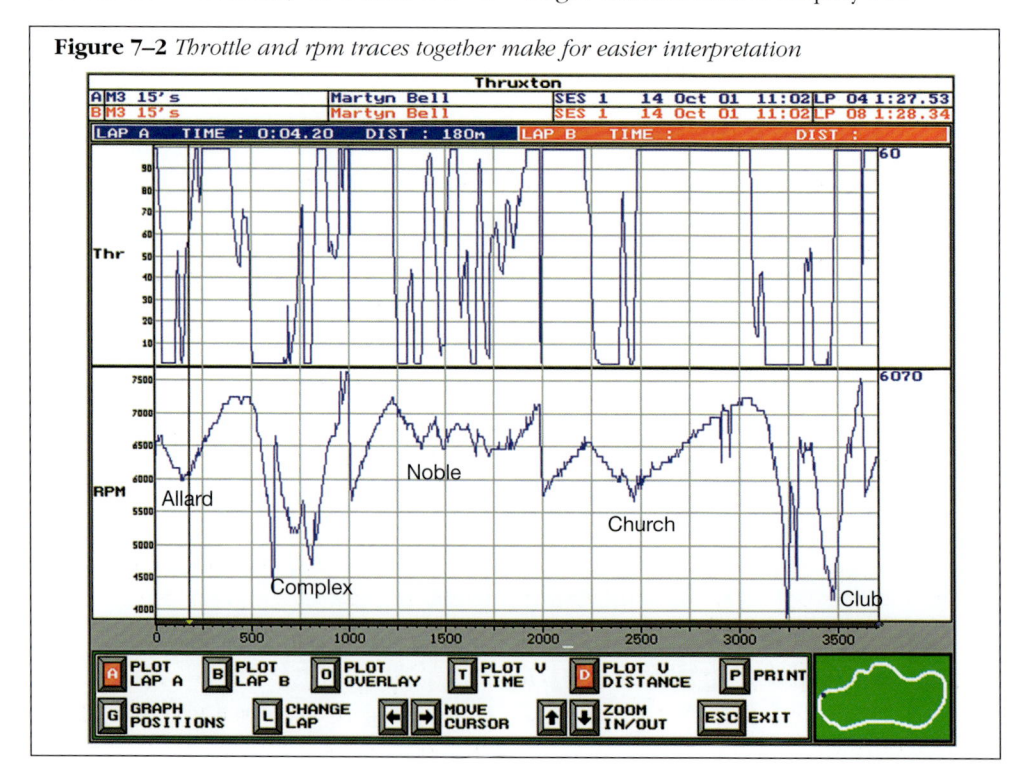

Figure 7–2 *Throttle and rpm traces together make for easier interpretation*

The throttle just reaches 100 per cent accelerating out of Allard when, at the 225m point, again Martyn backs off briefly, then gets hard on the gas as he powers towards the Complex. Again, the reason for the momentary lift could make you ask 'why?' For the next 140m the throttle is wide open and the revs climb accordingly. Slowing for the first bend in the Complex shows a part lift, a reapplication, and then a complete lift off the throttle. This would surely prompt another query. Remember, there are probably very good reasons for these occurrences, but as all of them relate to how fast the car is going as distinct (possibly) from how fast it could go if the throttle was open for longer, they're important details.

During the time the car slows for the complex, the rpm trace shows a sharp upward jump indicative of a downshift, yet the throttle trace stays flat at zero. We can deduce from this that Martyn doesn't use the hell-and-toe throttle blip on this downshift at least. The pattern is repeated on this lap and on others, so it's fair to conclude that it's a technique Martyn doesn't use.

You can then follow the rpm and throttle plots through the Complex, then at the apex of and just after Seagrave, the last of the Complex bends, there are two details to be picked out. The first we have talked about already where the rpm peaks suddenly and then comes back again, and this is followed rapidly by an upshift to fourth gear. We'll come back to that first detail again shortly.

The throttle is then wide open towards Noble, and then for the difficult fast left-hander itself, and the equally difficult right-hander at Goodwood you can see that the throttle mirrors the rpm trace – it really is an on-off section here, and again would warrant discussion with the driver to get his feedback on what was happening in the car, and why.

As the car exits Goodwood there is a clean upshift to fifth (top) gear and then its hard on the gas towards Church. This produces another interesting throttle trace, with Martyn lifting then briefly applying the throttle, then coming off it again before flooring it on the exit of Church. These details are only barely discernible in the rpm graph – the throttle trace makes it much clearer what is happening here, although just why would once more require the driver's feedback.

It's then flat out towards the braking zone for the Club Chicane. Here, what we previously thought of as slowing without braking followed by rapid slowing with the brakes is now evidently something slightly different. For at the 3,100m point, just after having initially lifted to slow for the Chicane, Martyn briefly reapplies the throttle, and only after that do the revs drop rapidly, indicating braking. Why did he reapply the throttle here? That's another query that could lead to some interesting feedback.

You can then see two downshifts, again without throttle blips, the short burst on the throttle towards the slow left-hander, followed by rapid slowing off-throttle for the slowest part of the Chicane, and then the rapid application of the throttle after the Chicane and the upshift to fourth just before the end of the lap.

So just by running quickly through a single lap we have raised a number of queries about Martyn's use of the throttle in his BMW. Let's zoom in on the early part of the lap to see if the occurrences there are any easier to interpret. Figure 7-3 shows the section from the exit of Allard Corner to the entry to Noble, including the whole of the Campbell–Cobb–Segrave Complex.

The partial lift at 225m is now more evident, and although it is short-lived, one has to ask why did this happen? Was it just a 'confidence lift' to settle the car? Did Martyn come across a slower car and have to hesitate briefly? Was the car right at the edge of its available grip and was the brief lift necessary to keep it on line?

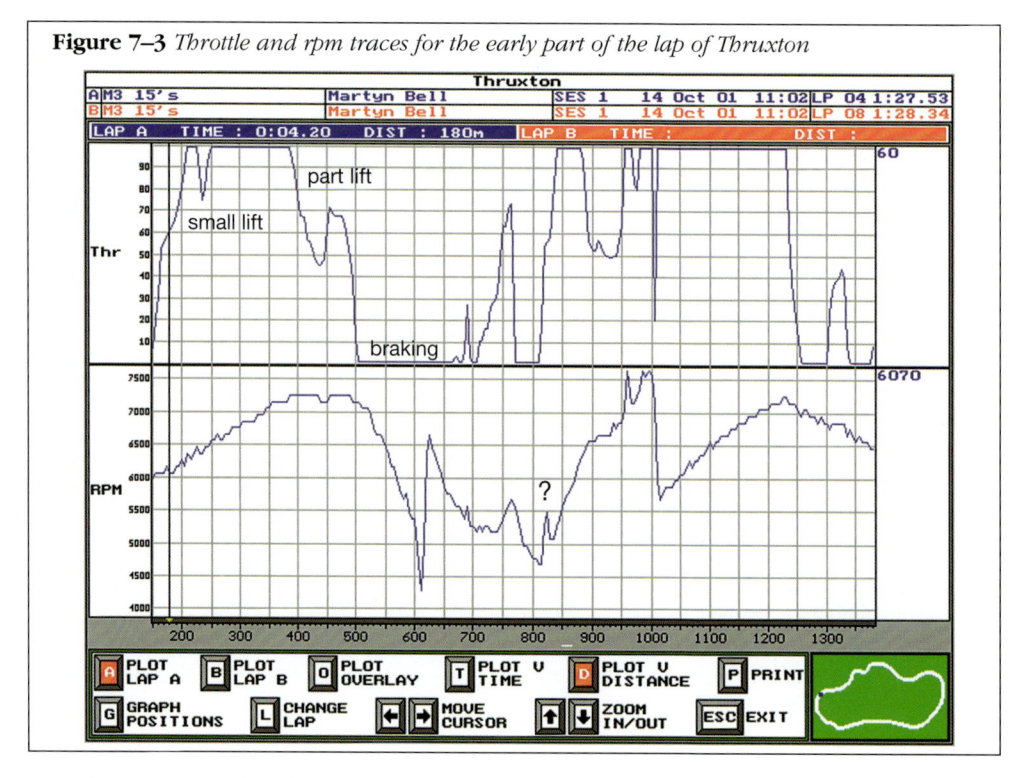

Figure 7-3 *Throttle and rpm traces for the early part of the lap of Thruxton*

Was the car actually at the outer limit of the road on the corner exit? At the moment all of these could be possibilities – driver feedback would be needed to find the actual answer (although when we look at lateral G-force in Chapter 9 we may gain more clues).

Moving on then, at 390m Martyn takes his foot off the throttle on the approach to the Complex. The partial lift followed by brief reapplication described above is more evident at this zoom level, and from its duration, and knowing that the track continues curving left prior to the right-hander at Campbell, we might surmise that the car is on something of a knife-edge here and acceleration needs to be checked, well ahead of the braking zone for Campbell.

At 820m there is another short-lived event that is barely discernible on the full-lap trace. The rpm trace shows a brief, rapid rise followed by a return to the expected rising slope immediately after the slowest point of Cobb Corner, indicated by the question mark. This is exactly matched by a hesitation on the throttle trace where the basically fairly rapid throttle opening is delayed for about 0.4 seconds. The likely causes? Two possibilities spring to mind from this pattern – either the car began to over-steer and lose rear-end grip, or (as we now know to be the case) Martyn ran the kerb here, releasing the inside rear wheel which prompted the brief rise in revs and he checked his throttle application, equally briefly, in response. Did he need to hesitate on the throttle? Was he just being kind to his engine and transmission? Or should he have stayed off the kerb in the first place so that the hesitation became unnecessary (which might have been altogether kinder to his car!)? Checking other laps to see if this happened more or less consistently might increase our confidence in the diagnosis, and also enable us to see the impact of the hesitation on rpm rise, that is, acceleration.

Running kerbs may well show up on the throttle and rpm traces.

Figure 7-4 shows two successive laps for this part of the course, the same two laps we looked at earlier in fact. You can see that at the 820 metre point the throttle and rpm traces are slightly different. In fact, the graphs show the car carrying more rpm through the apex of Cobb on the 'red' lap, and the throttle trace is rising slightly less steeply on the red trace (actually the slower lap overall) too. Thus, a smoother, neater approach should have yielded a small gain on the

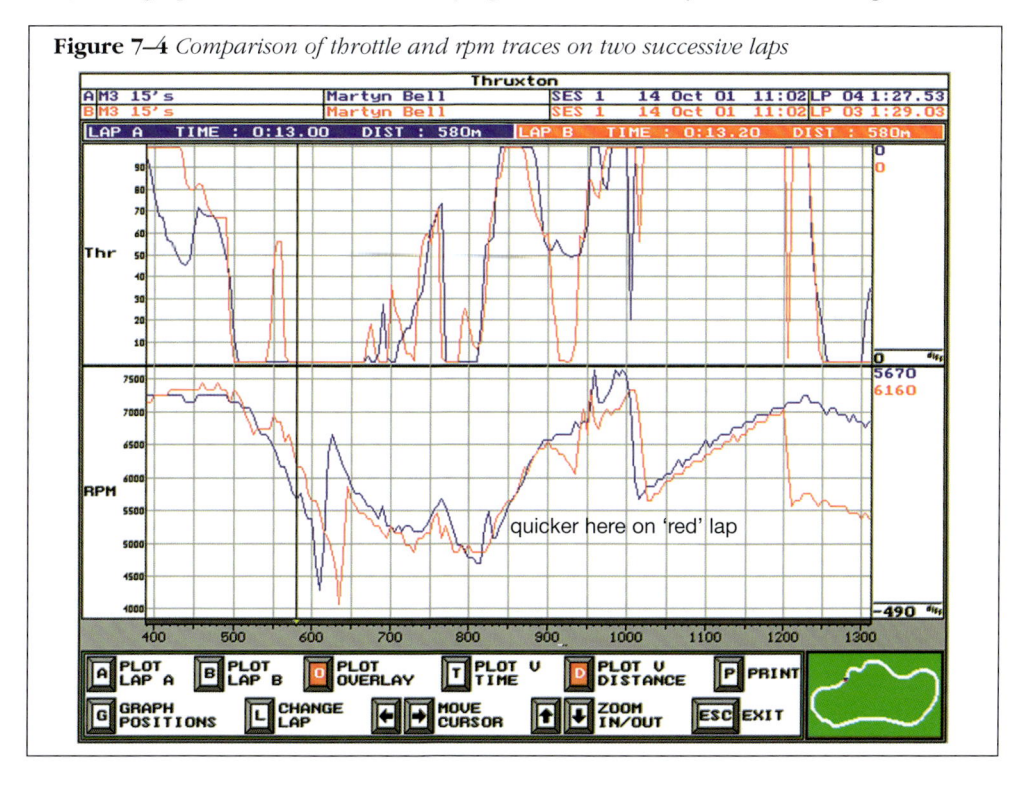

Figure 7–4 *Comparison of throttle and rpm traces on two successive laps*

red lap over this short section of the course, and analysis of the time difference between the two laps over this short section of the course showed that a tenth of a second was in fact made up on the red lap here.

We could go on examining all areas of the course where there were differences in use of the throttle, but hopefully these examples serve to illustrate some of the types of differences to look for, and how to look for them. As with a lot of the useful things you can discover with data logging, the real value is in the detail. It is also in asking why the throttle is at a particular position in any area of the track. There are times when you expect full, or zero or partial throttle, depending where the car is on the track. Any variations in that pattern, or variations from lap to lap, or run to run, are worthy of closer examination. They may help you determine why the car was quicker or slower on any one occasion, and if you know this then you will hopefully be able to ensure the car and driver go as quickly as possible everywhere! If only it was that simple . . .!

In lapping Thruxton with our kind demonstrator/instructor Martyn Bell we have been able to see some of the nuances to look out for. Clearly some of these relate to what one might call driver behaviour, whilst others might be more closely linked to chassis behaviour. In all cases though, since the throttle is a manually operated control, by examining the throttle position trace we are looking at the driver's responses to the inputs he detects. Then, by asking the driver questions (or by asking yourself if you are the driver and data analyser) you will possibly be able to find out if that tail-out moment that required a big throttle lift was the driver being over-eager on the gas pedal, or whether the car is set up in a tail-happy condition. As professional driver coach Nigel Greensall told me, 'some people are too aggressive on the throttle, maybe too soon in the corner. Maybe they've lost too much speed on the entry to the corner, and then they use too much power too soon, which then causes more problems.'

But how do you tell if a driver is applying the throttle too soon or too late coming out of a corner? This is something that depends on the car, where in the corner it is (in other words, how much of its tractive capacity is already in use cornering the car), and in particular how much power it has. Assuming equal levels of aerodynamic downforce, a powerful car can break traction far more easily than a less powerful car, and can do so in fast corners as well as slow corners. It is therefore impossible, I would say, to give hard and fast rules on when the throttle should be applied. Suffice it to say that it is unlikely that full throttle can be applied at the same time a car is cornering hard unless it is a very low powered car, or one running very high downforce – tyres cannot exert high accelerative forces at the same time as they exert maximum cornering grip. (The same is true of braking, which we shall look at later.)

However, if you can view the data of a quicker driver against a slower driver in the same car, and you can see that the quicker driver gets on the throttle, and up to full throttle, sooner than the slower driver, then the slower driver can learn easily what he needs to do to go quicker in those particular parts of the course.

Even without another driver to compare to though, instances like one or two of those we have looked at from these laps of Thruxton can still make you question whether different throttle technique might make up some time, and comparisons with one driver on different laps can prove the point. When used with rpm and speed too, it is also evident when the use of the throttle can yield gains.

When I asked Nigel Greensall: 'What if you're looking for the symptoms of understeer and oversteer, can you diagnose those categorically from the data or does that have to tie up with driver feedback?' his reply was quite thought-provoking. He said: 'That's difficult really because handling characteristics of cars are almost always induced by the driver. And so whichever way you tune a car, if you put a different driver in it he can produce different handling characteristics. At the extremes you can put driver A in the car and he'll produce understeer and put driver B in and he'll produce oversteer, and that's not related necessarily to how the chassis is set up. But where the data logging can be useful is in actually understanding why driver A understeers through a corner and why driver B oversteers through the corner.'

So, can throttle position, allied to the other channels we've looked at, be used to point to particular chassis characteristics? Yes and no! Which is to say, it can raise questions as we have seen, and by talking with the driver you might be able to connect particular types of trace to particular handling characteristics. The easiest condition to spot is loss of traction. This may happen in a straight line or coming off a corner, and the symptoms to look for in the data traces are the same as those the driver feels – a rapid increase in revs accompanied by a speed increase that may be reduced from 'normal', and the possible backing-off of the throttle to contain the wheelspin. This condition may produce similar symptoms to oversteer on the channels we have looked at so far. To be able to diagnose that condition, or its opposite, understeer, would really need the benefit of steering position logging, which we shall look at next, and lateral G logging, which we shall come to after that.

Another useful facility of DAS analysis software that we can use relating to throttle position, mentioned earlier in this chapter but which is pertinent to this particular section, enables the reporting of histograms of throttle position through a lap, or through a pair of laps to enable comparison. At the start of this section I said that a car would generally set quicker times if the throttle spent more time being fully open, and this type of histogram allows you to measure just that. Figure 7-5 is such a histogram, and it

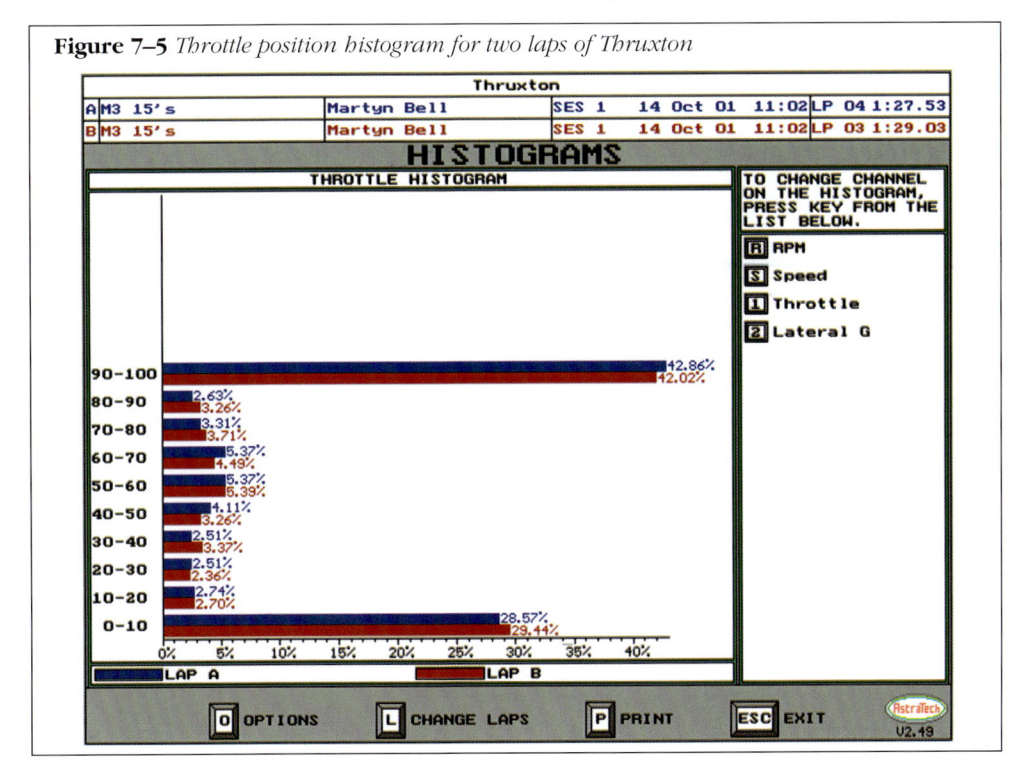

Figure 7–5 *Throttle position histogram for two laps of Thruxton*

shows the percentage of time through the two laps we have been comparing that the throttle spent in each of the 'brackets' indicated on the left.

Clearly, the throttle spends most of its time (over 70 per cent) either fully or nearly fully open, or completely or nearly completely closed, with all the other brackets making up the balance of the lap between them. An interesting detail however is that on the quicker lap the throttle was wide open for 0.84 per cent more of the lap, and closed for 0.87 per cent less of the lap. This may be a coincidence of course, but a check on some other laps reveals the same pattern so it is more likely that there is a direct correlation. Naturally the actual percentage of the lap for which the throttle is wide open will depend on the circuit, the car and the driver.

Throttle position is thus a revealing and fairly simple to interpret channel, showing the use of one of the main vehicle controls. The next chapter looks at another main control – the steering.

The throttle trace can be very revealing . . .

Chapter 8

Steering angle

AS WITH THE other channels we've looked at so far, let's start with the basics. When looking at steering angle, what is being measured and logged, and how does it appear on the analysis screen? Steering angle is the measurement of the rotational displacement of the steering from the straight-ahead position. The greater the deflection of the trace from the 'central' straight-running position, the more the steering wheel has been turned.

The steering angle sensor detects the rotation of the steering and outputs a proportional voltage to the DAS. If possible, calibrate your steering channel to relate to steering wheel angle, not front wheel angle or any other abstract quantity. It's the steering wheel that the driver connects to, after all.

Just like the throttle pedal, the steering wheel is one of the most important links between the driver and the car, and the

Calibrate your steering channel to relate to steering wheel angle if possible.

86 COMPETITION CAR DATA LOGGING

Figure 8–1 *Steering angle plotted with speed for a lap of Donington Park*

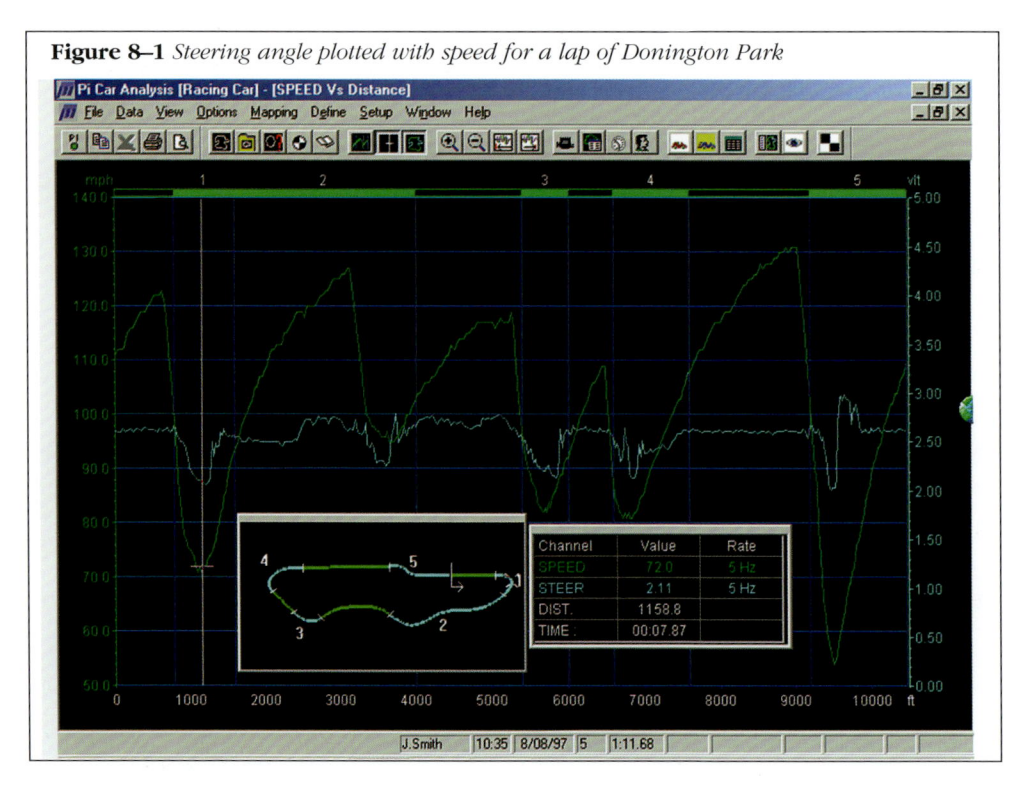

Figure 8–2 *Donington Park 'National' circuit, 1.957 miles*

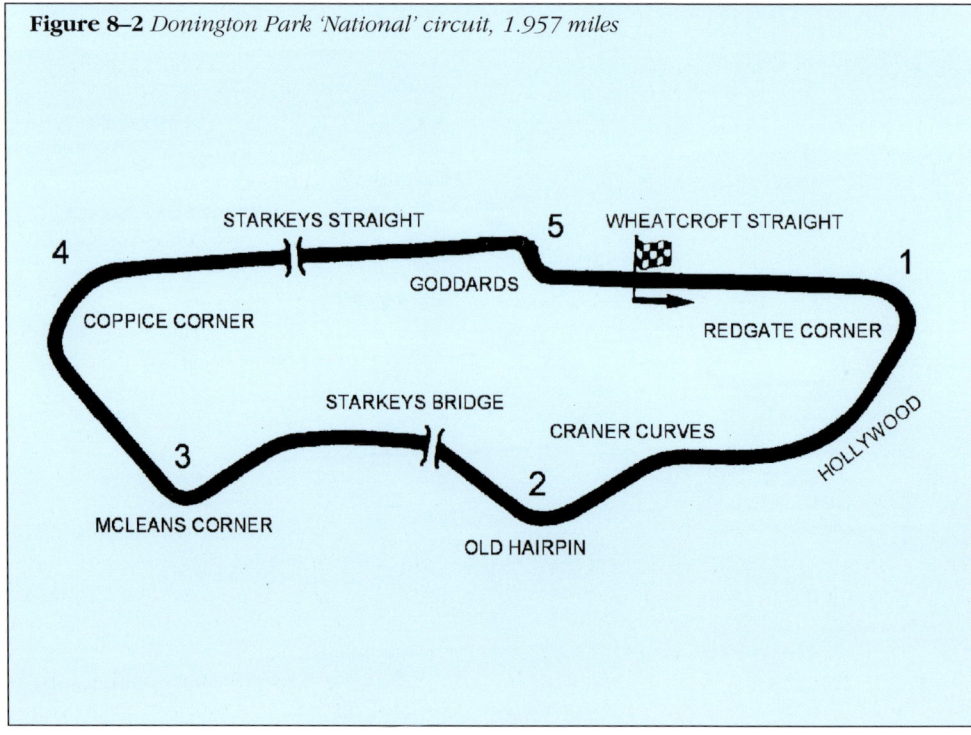

inputs and responses on the steering are fundamental to diagnosing all sorts of events. Equally, the inter-relationships with other logged channels are important too. Mike Holmes at Pi Research has some good advice here: 'Never look at the steering without the throttle.' In other words, examine these two channels together to see, often at a glance, how the driver has behaved or responded in a given corner.

But first, in order to display an easy-to-interpret steering trace, Figure 8-1 shows just steering with speed. The track this time is the 1.957-mile National circuit at Donington Park in Leicestershire, England (see Figure 8-2 for an actual track map.) The car was a Formula Vauxhall Lotus single seater (2 litres, 180bhp, control slick tyres, simple wings), and the data was logged by a Pi Research system. The pale blue line is the steering trace, with the left-hand vertical axis displaying the sensor output in volts from 0 to 5 (not ideal, but it does show semi-quantitatively what is happening to the steering). The green trace is speed, and the track map helps to see the various corners where the minimum speeds occur.

The steering trace essentially follows the speed trace because all the principle corners are right-handed. Hence, the downward dips in the steering represent the application of right lock. This is easy to remember if you think about looking along the steering trace from the left-hand end of the graph; deflections to the right equate to right turns, and vice-versa.

The vertical white cursor line has been positioned at the slowest point of Redgate Corner, the first right-hander after the start/finish line. You can see on the steering trace that right lock was applied before the slowest point in the corner, and was then wound off again as the car accelerated out of the corner. All very neat and tidy – on this lap! The next right-hander, preceded by a fast left curve for which the speed doesn't reduce, is faster, and shows a similar steering pattern. There is a small spike in the trace as the car starts to increase speed exiting the corner. This shows the brief application of opposite lock, or more accurately a rapid correction in which the applied lock was wound off. More on this shortly, but already you can see that events about which we were earlier surmising by studying rpm and throttle traces become a lot clearer if you can see what was happening with the steering.

The straights are very evident on the steering trace – the line hovers at or near the mid-point in the measuring range. Notice that the dead-ahead position is actually at about 2.6V on the steering sensor output, not 2.5V which is actually the middle of the 5V range. This is not a problem as such, it just shows that either the sensor or the steering wheel was physically very slightly off-centre.

The left-handed curves on the track are less obvious initially, and show as upward deflections on the steering trace relative to the straight-ahead position. However, near the end of the lap the right/left chicane shows up very clearly with first the downward (right lock) deflection followed immediately by the upward (left lock) deflection.

So that's basic navigation with the steering angle trace. But as with the other channels we've looked at, and as was hinted at above when we mentioned the occurrence of a steering correction, the real value and interest is in the detail. Furthermore we'll now take up the earlier advice to study throttle angle and steering angle together, and we'll do so with a more eventful lap from the same session as the lap we've just looked at. The lap shown in Figure 8-3 was actually the next lap in this session, but the steering trace has lots of interesting details to examine. The steering output scale has also been expanded somewhat by resetting the maximum and minimum values displayed to 3.50V and 1.50V respectively via the software using the 'scale graph' option in Pi Research's Club Expert analysis software (found under 'Options' or by right-

Figure 8–3 *Steering details*

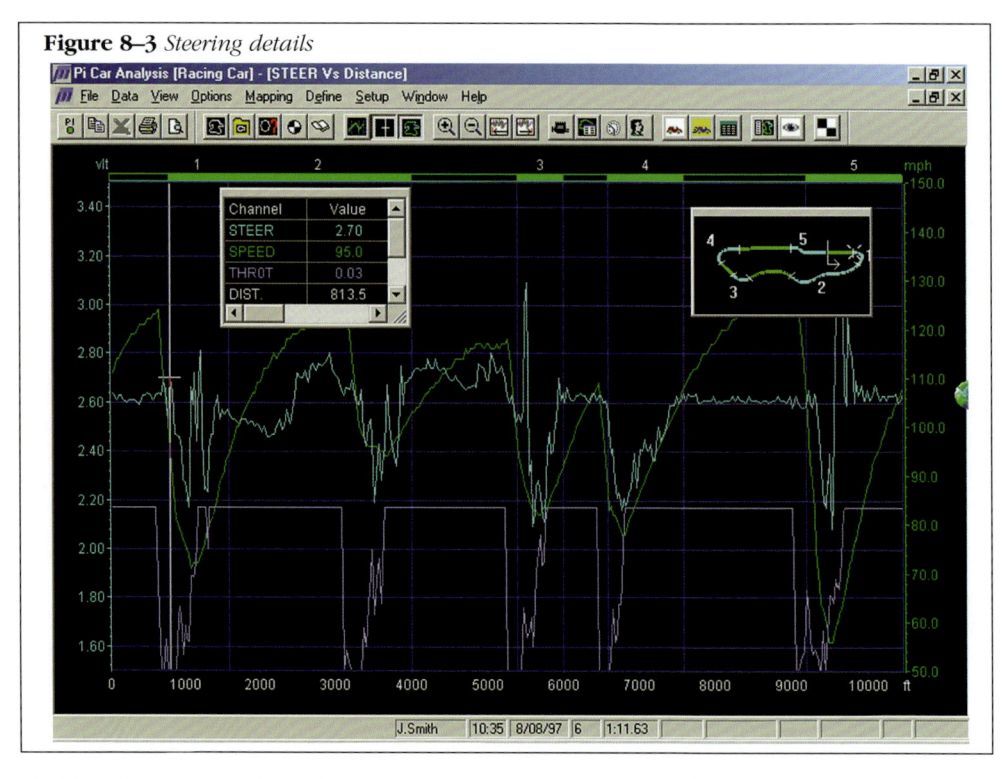

clicking the mouse when the pointer is in the graph body). See Appendix A if this sounds like technobabble!

You can see immediately that the steering trace at the first corner, Redgate, is very different to the previous lap. Instead of a simple downward curve that followed a similar shape to the speed trace, the steering shows the initial application of right lock being followed by two upward spikes, representing two rapid doses of opposite lock. And this time the steering really was wound far enough to call it opposite lock since left lock was briefly applied during these corrections.

But let's back-track slightly because there is even more interesting detail here. The speed and throttle traces show where the driver began slowing for the corner, at around 620 feet (as the distance units are displayed) into the lap. The throttle trace drops to zero and the speed begins a rapid decline. At about 810ft a small amount of left lock is applied (presumably to take the car right to the outside of the track), and then a tiny amount of right lock is wound just prior to the entry to the corner, this while the brakes are still being firmly applied because the speed is still declining very rapidly.

At about 885ft the driver stops braking, as shown by the less steep decline in speed and at 930ft right steering lock is firmly applied to turn the car into the corner. At the same instant the driver also applies a small amount of throttle, but as the speed is still declining this looks like a last-minute downshift on the entry to the corner (the rpm trace, not shown here, seems to bear this out, there having been two other downshifts prior to this).

Then the fun starts! At the point where maximum right lock has been applied the driver is already increasing the throttle opening, and in response to this two things happen; first and least surprisingly the speed starts to increase, and second, the driver very rapidly adds a steering correction resulting in the return of the steering wheel to the dead-ahead position, then back to slight right lock, and then, as the throttle is fully opened to opposite lock, that is, slight left lock. The corner is then finished with right lock, which is gradually wound off as the corner opens out and the car heads off to the next challenging piece of track.

Clearly the driver was having a much more aggressive go at this corner on this lap than on his previous lap. He was very early on to the throttle and this resulted in two quite big steering corrections. The driver was also determined to stay on the throttle, as shown not only by how soon in the corner it was up to full throttle, but the fact that he barely hesitated on the gas during the first steering correction, and kept it nailed during the second.

From the sidelines this will have looked pretty impressive, but by trying to get on the throttle so soon and so hard, was the driver being a little unruly? The 'split report' which shows the segment times actually showed on this lap that this corner was taken seven hundredths quicker than the previous, tidier lap. But maybe a tidier compromise would be faster still? We'll come back to this point in the next chapter when we look at lateral G . . .

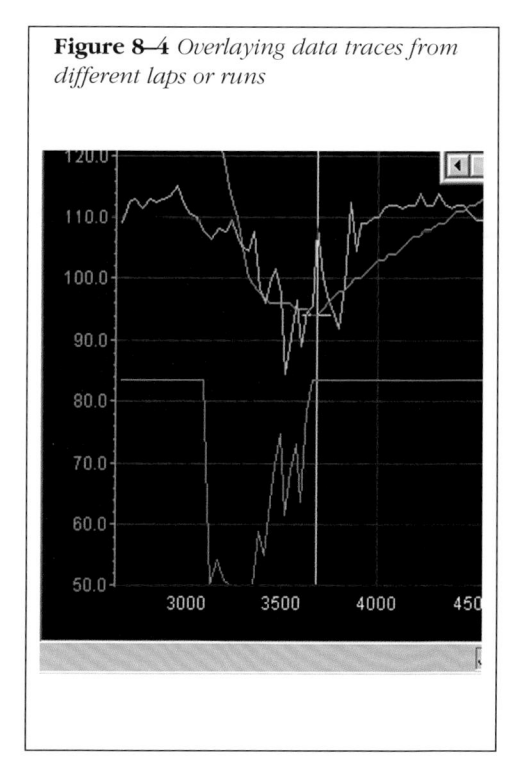

Following Redgate the car proceeds on full throttle through the right-handed curve that is Hollywood Bend, then plunges down the hill through the demanding Craner Curves. This is then followed by Old Hairpin where the drivers have the job of slowing their cars from a fast downhill section for this fairly sharp corner. In this case speed peaked at 128mph coming after Craner Curves and dropped to 94mph in Old Hairpin. The traces at the entry to the corner are interesting, and figure 8-4 shows this section zoomed in to reveal the detail with a little more clarity.

The driver comes off the throttle and starts to brake (the speed reduces rapidly) at around 3075 feet, and he has also driven a line that allows him to straighten the steering, more or less, at the same instant. The small throttle blip shows where he heel-and-toe downshifted to the next lowest gear, and then at about 3,341ft he comes off the brake (the speed decline reduces in rate) and he adds right steering lock to initiate turn-in. He also gets back on the throttle at this point too. But, he next reduces steering lock and throttle, then tries again to apply considerably more throttle only to apply even more lock and back off the throttle at the same time (this at the 3,500ft point). This process is repeated again, with throttle application being followed by an increase in right steering lock and backing off the gas again.

Does this sound familiar? Could the driver be struggling to get the car to turn the corner? With the downhill approach and braking area it wouldn't be a surprise. The speed trace also tells a tale through this corner too. Cast your mind back to Chapter 4 when we looked at a corner on a hillclimb that tended to provoke understeer, and we discussed the way the rpm trace (in relation to speed)

Figure 8-4 *Overlaying data traces from different laps or runs*

got 'stuck' in a downward sloping valley. Well the speed trace here does the same, showing that the driver's attempts to accelerate the car were not working. Here though we have confirmation that understeer was the problem with the way the steering and throttle traces interact.

Finally, at about 3,600ft the throttle goes wide open and corrective lock has to be applied as the driver balances the car by breaking the back end loose on the throttle, and then the corner is finished off with right lock and one more small opposite lock correction. Looking over the whole corner, it appears that the car exhibited corner-entry and mid-corner understeer, and this was finally cancelled out by the driver with aggressive throttle application only at the corner exit. Earlier attempts to get on the throttle were rewarded with the car pushing on, requiring an increase in lock and throttling back to try to get the front to grip again.

Steering data can in fact be an indicator to various handling problems, especially in association with the other channels we've looked at. As we've seen here, rapid and brief reductions in steering angle, or the application of actual opposite lock during cornering, which may or may not accompany a throttle lift, can indicate oversteer. The steering inputs will be fairly obvious, but how the driver reacts or indeed initiates the trait with the throttle will depend on many things, not the least of which will be whether the car is front or rear-wheel drive. A front-wheel-drive car may be provoked into oversteer by lifting off the throttle in a corner, and having done so is unlikely to respond well if the driver then lifts off the throttle further – the usual result of so doing is to complete the spin. The preferred response is usually to get back on the throttle which reduces the lateral grip at the front end, and so balances the car again, and this would be apparent in the throttle trace. With a rear-wheel-drive car, as we have seen the application of corrective lock may or may not be in company with a throttle lift, and that will depend on how powerful the car is, how well balanced it is and how aggressive, or perhaps we should say assertive, the driver is.

Understeer on corner entry can be determined as we have seen from the way the driver feeds in steering and throttle inputs, and mid-corner and corner-exit understeer would be accompanied by frustrated attempts to get on full throttle until very late in the corner. Sometimes the condition can be cancelled out by the driver treading hard on the throttle, but this can make things worse.

These factors depend again on which end the car drives through, and also the balance and set-up of the chassis, and the geography and topography of the track. Clearly an understeering front-wheel-drive car is unlikely to be balanced by using more throttle, and in this case we would expect to see a steering trace not unlike the one in the first part of the corner we have just been studying. The driver would quite likely be trying to increase his throttle angle to accelerate through and out of the corner, but he may have to keep backing off and applying more or less steering to keep on the road. A clever driver in a front-wheel-drive car in that situation would probably do better to turn in quite late and quite aggressively, but stay off the throttle until late in the corner, thus giving his front tyres more chance to generate cornering grip through the bulk of the corner. The data traces would show few steering corrections through the corner and later throttle application, but the segment time might be quicker.

We have said previously that it is important to compare different laps to see if a trace pattern is consistently repeated before deciding on a diagnosis, and the same applies to steering. As we have seen, the two consecutive laps we have looked at here showed very different events just in the first corner of the

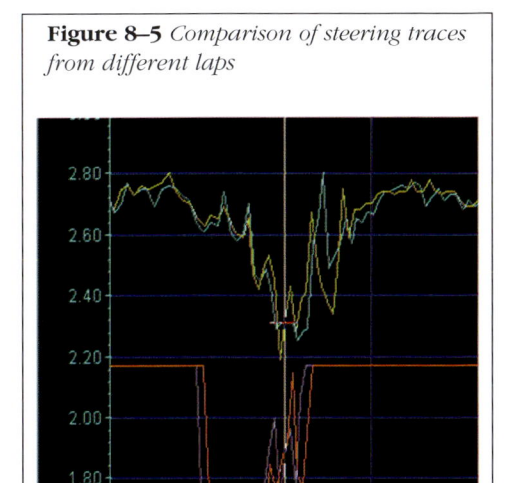

Figure 8–5 *Comparison of steering traces from different laps*

Figure 8–6 *Steering and throttle comparison at McLeans Corner on two different laps*

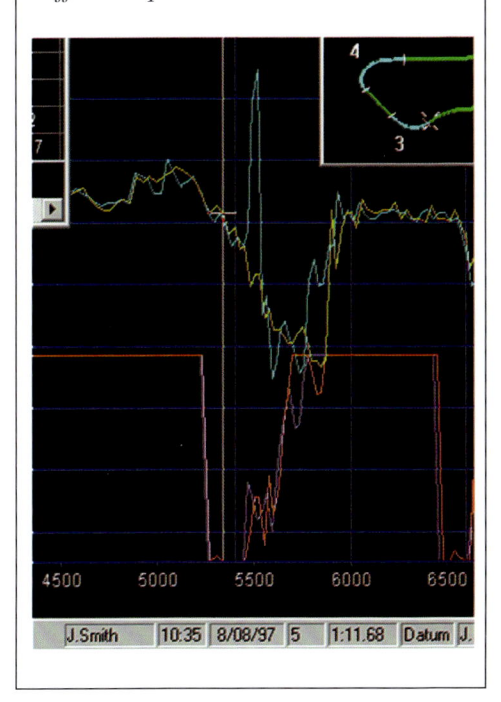

track. This tends to point at differences in driver technique (in this case he was apparently trying harder on the later of the two laps) rather than chassis imbalances. However, without querying this with the driver, how can we be sure he wasn't driving in a way that overcame a chassis problem? Data logging is a tool that provides many questions, but it cannot supply all the answers.

Looking at the data from the Old Hairpin on the two laps we've been studying here reveals striking similarities in both the steering and throttle traces. Figure 8-5 illustrates this and here the yellow steering trace and the orange throttle trace are from lap 6, whilst the blue steering and purple throttle traces are from lap 5. The white cursor has been placed in the middle of the corner, so to speak. The two pairs of traces follow each other really very closely until quite late in the corner. Even then the differences appear only to be a matter of timing in the application of throttle and steering adjustments. So the driver's inputs and responses were consistent over these two laps in this corner.

Move on to the next corner, McLeans, however and the steering traces look vastly different (Figure 8-6). Shortly after turn-in on lap 5 the steering trace shows a big dose of opposite lock being applied which matches up to the throttle being opened fairly early in the corner. This is completely absent from the trace of the previous lap. But the previous lap does show that getting on to full throttle too early in the corner exit phase was associated with having to back off again while at the same time winding on more steering lock. Thus, the driver appears to have induced corner exit understeer on lap 5, and just maybe that's what prompted the very early throttle application on lap 6. The driver may have been attempting to prevent that onset of understeer, but in so doing created an oversteer event early in the corner.

As stated at the beginning of this chapter, the steering trace can be highly illuminating. Diagnosing the causes of understeer and oversteer is not really within the scope of this book, but hopefully we're beginning to spot some of the symptoms by using data. Next we'll look at lateral G and its interpretation. This can help even further with spotting handling imbalances.

Bet that showed up on the steering trace . . . and all the other channels!

Chapter 9

Lateral G

LATERAL G IS essentially a measure of cornering force, or more strictly, cornering acceleration, and for convenience is usually expressed in units of G where 1G is equal to the acceleration due to gravity on planet Earth. Expressing cornering as an acceleration is sometimes confusing. You might travel around a large radius bend at a steady speed, yet you are accelerating. How so? Well, in order for your car to turn, you rotate the steering wheel so that the tyres generate sideways grip to provide the force to make the turn happen. If they didn't generate a sideways force the car wouldn't turn, it would continue in a straight line. This is Newton's First Law of Motion in action.

When our tyres generate a turning force, Newton's Second Law of Motion says that our car will accelerate in proportion to the force acting upon it, and also in proportion to the car's mass. The turning force is actually directed towards the centre of the turn, and so we can say that the car feels an acceleration in the same direction, that is, towards the centre of the turn. In physics it's called 'centripetal acceleration', which means 'centre-seeking', but we call it lateral acceleration for the simple reason that it is directed laterally in relation to the car's intended direction of travel. We also tend to use 'cornering force' as synonymous with 'lateral acceleration'.

Cornering force acts towards the centre of the turn.

Searching for the limiting speed...

Don't confuse this with centrifugal force. Centrifugal, or 'centre-fleeing' force is only an apparent force caused by inertia. It manifests itself in ways that make it seem as though a force is being exerted, such as when your head tries to flop towards the outside of the turns. But the fact is, the turning force, and hence the lateral acceleration, is directed towards the turn centre. There, that's another bee in my bonnet sorted out!

It was mentioned above that lateral acceleration is usually stated in units of G where 1G is the acceleration, or force, due to gravity. This is easier than quoting so many 'metres per second per second', or 'metres per second squared' as the convention is sometimes expressed. The acceleration due to gravity is approximately 9.81 metres per second per second, or 32.2ft per second per second, but calling it 1G saves time, ink and paper.

The cornering acceleration that can be generated by a competition car (or any other car) depends on the grip available from the tyres, and also on the vertical force acting on the tyres. Soft compound tyres generate more grip, and enable higher cornering Gs. Aerodynamic downforce can increase the vertical force on the tyres without actually increasing the weight of the car, and it enables higher grip and hence higher cornering Gs at speeds where significant downforce can be generated.

However, in performance terms the lateral acceleration in a corner is related to two basic parameters; the speed through the turn and the radius of that turn. To be exact about it, the lateral acceleration is equal to the square of the speed divided by the radius of the turn. So, a higher speed in a given turn generates a higher lateral G. Clearly, if a car generates a certain amount of grip and downforce at a certain speed, there is a limiting speed at which a given radius turn can be negotiated.

This is 'the limit' that every competition car driver searches for and claims to be at the whole time! If you fail to reach the limit you will be slower than you could be, exceed it and you'll fall off the road.

This is partly what makes the study of lateral Gs, and the driving of competition cars, so interesting. You can determine, by talking to your tyre supplier, what levels of grip (coefficient of friction) your tyres should generate, and in simplistic terms this tells you what your maximum lateral Gs could be. So say the coefficient of friction of your tyres is 1.4, then expect to achieve roughly 1.4G peaks in the turns. Of course, this depends on the track surface and conditions, chassis set-up, and driver skill and experience. And if your car generates significant downforce, this too must be taken into account by multiplying in an additional 'grip factor' that is related to speed. (Downforce increases with the square of speed; see my book *Competition Car Downforce* for the full explanation of that aspect.)

Let's now take a look at a lateral G trace (Figure 9-1) plotted with speed for ease of recognising position on track, from the Donington Park data we looked at in the previous chapter. Remember this was set by a Formula Vauxhall Lotus single seater. Lateral G is in red with speed in green again. You'll see that Pi Research refer to lateral G as 'ACCEL' here, and on other systems you might see it variously referred to as LatG, LtG, G Lat and LATAC.

The first thing to notice is that there are up and down lateral G peaks, and these correspond with the left curves and the right turns of the track respectively. Let's run quickly through the lap again to get our bearings, so to speak. Redgate Corner sees the first dip in the speed trace at about the 1,100ft point. The speed falls to a minimum of 71mph, and the G trace shows a maximum of -1.71G (minus, -, for right turns, plus, +, for left turns).

The speed then builds after Redgate and the G falls away but then climbs to a lesser peak in Hollywood Bend (2,160ft) of 1.17G when the speed is 114mph. We know (from the previous chapter) that

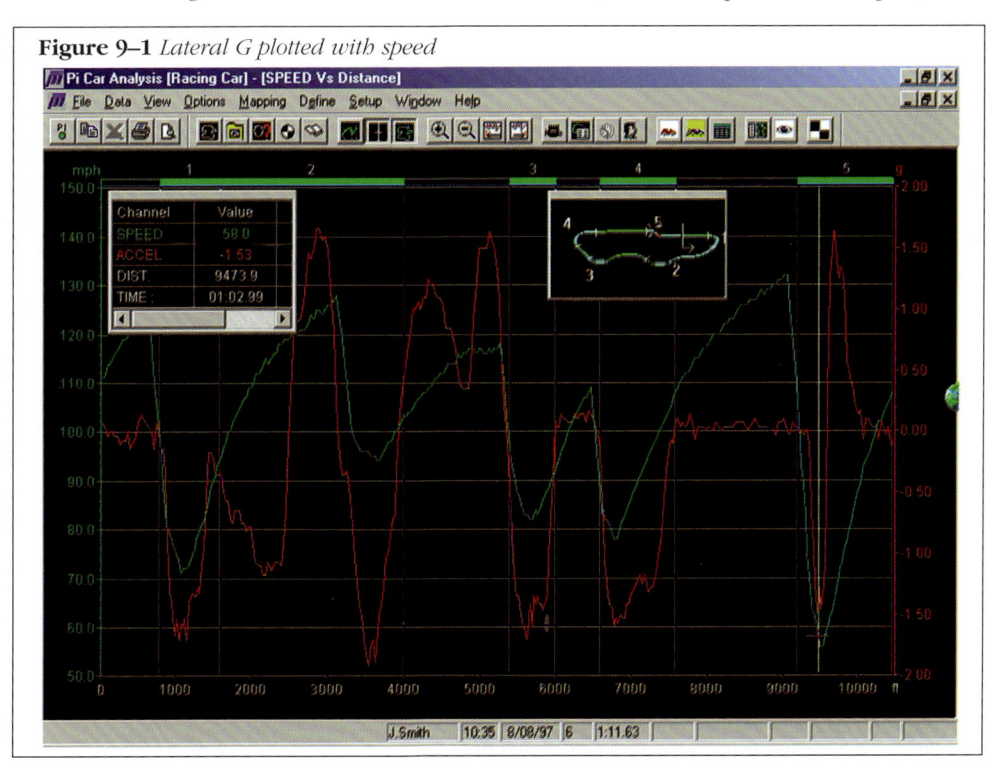

Figure 9–1 *Lateral G plotted with speed*

the driver is on full throttle here, and the G peak is obviously much lower than the others around the track, so this is not a significant corner in this car. This is an important point to keep in mind – it's not usually worthwhile stopping to analyse the data at a point like this. The driver is hard on the gas, the car is not heavily loaded at this G level, so what more is there to say?

The left-handed downhill sweep that is Craner Curves sees a G peak of 1.67G at 125mph at 2,900ft, just before the speed reaches its maximum on this part of the course, and we know the driver was still flat on the throttle here too.

Into Old Hairpin at 3,600ft the speed drops to 94mph but lateral G peaks at 1.92G, the highest seen around the lap. The remaining bends and curves in the lap show up clearly on the trace, as does Starkey's Straight (a more or less flat part of the trace) and finally there's the right–left chicane just before the finish.

We could begin to speculate about the differing G levels reached in each corner. For example, why is the maximum G at Old Hairpin so much higher than the other corners? The first thing to check is that this is a consistent feature, and looking at other laps in this session, G peaks of up to 2.00 occurred here whilst other corners also 'only' showed the 1.6 to 1.7G levels seen on this lap. So this is more likely to be a factor relating to the track – a more grippy surface or some other track characteristic that helps the car generate more grip here.

So that's the generalities of lateral G. As with the sensors looked at previously, let's now study the details of some data more closely to see what interpretations we can make, and what diagnoses might speculatively arise from those interpretations. We have been building up our armoury of interpretive tools as we have added each sensor to the story so far. By now looking at all the sensor traces we can tell with far more certainty what is going on. Having said that, putting up too many traces on screen at once can be confusing, so we'll look at three or four at most at a time as we try to figure out what was happening to our Formula Vauxhall Lotus car around Donington Park. Cross referring to other traces is easily done with the DAS software though – you can pick which channels you want to display at any one time.

Oversteer

Figure 9-2 shows part of the same lap as we looked at in general above, but now zoomed in to show more detail. In particular, look at the area of the graph around the first speed dip (green trace) on the left, which corresponds to Redgate Corner. Just three traces are used here for clarity; speed, steering (blue) and lateral G (red). Running through the corner we see that as the driver turns into the corner with right lock, the speed is still falling, and the lateral G starts to build. At around 1,065ft, where the cursor is placed in fact, the lateral G peaks at –1.71G, and this also corresponds to maximum steering lock and minimum speed in this corner too.

This is immediately followed by a steering correction, with a brief fall in lateral G, then a reapplication of right lock

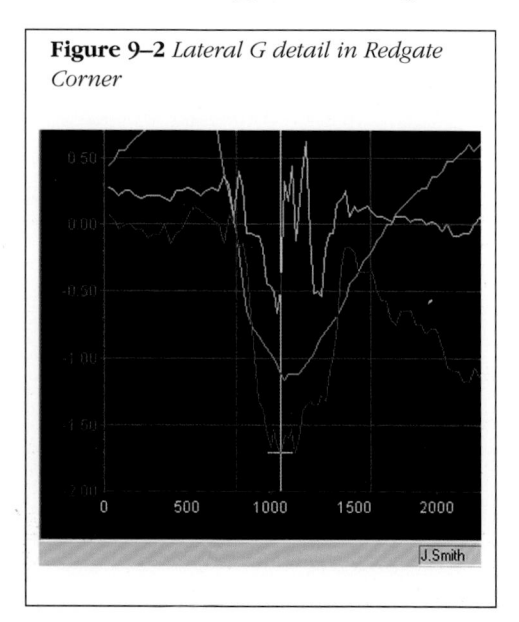

Figure 9–2 *Lateral G detail in Redgate Corner*

which prompts another lateral G peak of −1.71G at 1,150ft, and then a bigger steering correction which sees the lateral G tail off at around 1,215ft. We can also see that speed starts to climb, slowly at first but then more rapidly once the steering corrections are finished. From there, even though right lock is applied again, the lateral G tails off, then falls away completely in line with the steering being fed back to 'straight ahead'. Speed continues to climb.

What is interesting here is that we can see that maximum lateral G corresponds to the point where maximum lock and minimum speed occurred, which won't necessarily always be the case, but that the lateral G reduced as soon as corrective steering lock was applied. In other words, whatever the driver was doing here caused a reduction in cornering grip.

On the basis of these traces we could surmise that the driver was dealing with an amount of oversteer in this corner, which was why the steering corrections had to be made. So let's look at the throttle trace in conjunction with these traces to see what help that provides – see Figure 9-3. The throttle trace is in purple. Now we can see that the throttle was being opened very early in the corner,

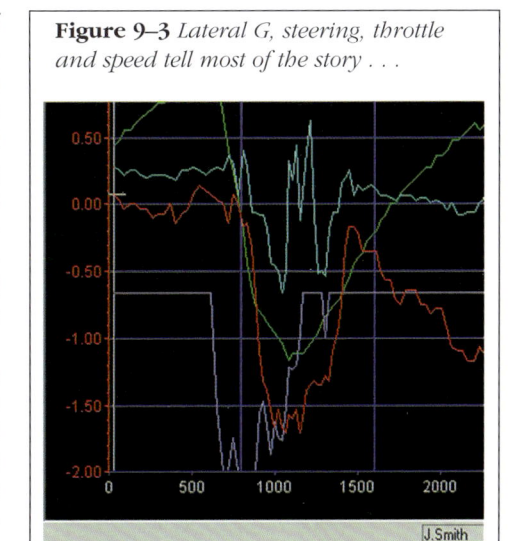

Figure 9–3 *Lateral G, steering, throttle and speed tell most of the story . . .*

just after the driver came off the brakes, going by the speed trace, but this looks as though it was just to maintain speed to the corner apex because only small throttle openings were being used.

The first interesting part of the throttle trace is at around 1,065ft, which as we have seen is the point of minimum speed, maximum lock and the first lateral G peak. Here the throttle is being opened rapidly, and this corresponds exactly with the first steering correction,

Rapid throttle opening can lead to corner exit oversteer. (Tracey Inglis)

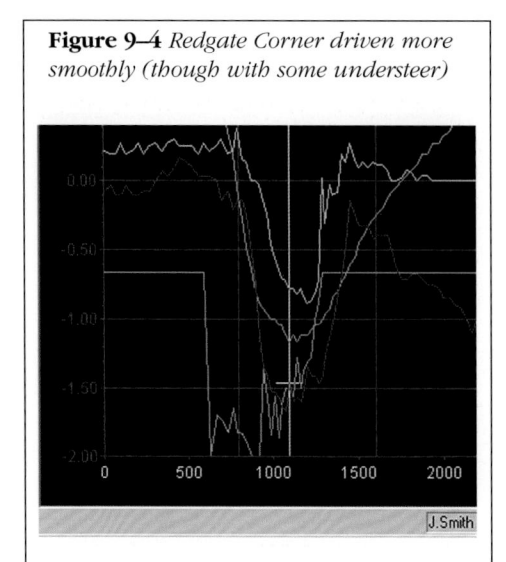

Figure 9–4 *Redgate Corner driven more smoothly (though with some understeer)*

and the small subsequent drop in lateral G. The throttle trace then shows a brief hesitation shown by the 'ledge' at 1,090 to 1,130ft, and here the steering lock is reapplied and the lateral G peaks again. The throttle then goes rapidly to fully open, which is followed by the bigger steering correction and the reduction in lateral G. However, there are no dips in the throttle, the driver merely applies corrective lock and stays on the gas. But clearly the steering corrections were necessary because of the throttle application.

By now the lateral G has declined although the corner is not yet over, and this may reflect a loss of grip caused by the oversteer which the driver induced in his enthusiasm to get on the throttle early! On the other hand, this may have been the best technique for getting this car through and out of this corner most efficiently. Certainly the lap displayed here produced the best split time for this corner in the 'split report', being a few hundredths quicker here than the smoother and more conservative use of the throttle shown in Figure 9-4. With a more powerful car or in slippery conditions this may not have been the case . . .

The single seater we have looked at in these examples seems to exhibit over- steer to varying degrees in most corners and on most laps. Usually it is evident in the corner-exit phase, but the example in Figure 9-5 at McLean's Corner shows the car oversteering on corner-entry. The cursor has been placed at about the point where the driver comes off the brakes and is just about to reapply the throttle. Lateral G has already started to build even though only a small amount of steering lock has been applied. There is then a large steering lock correction (left lock), which prompts a 'ledge' in the building lateral G trace, after which the car continues towards its maximum lateral G point. Moderate mid-corner oversteer is then displayed as the driver gets on the throttle and applies a more modest dose of opposite lock which sees the lateral G decline at the minimum speed point. The last phase of the corner seems to display a well-balanced exit as the throttle goes wide open and the steer-

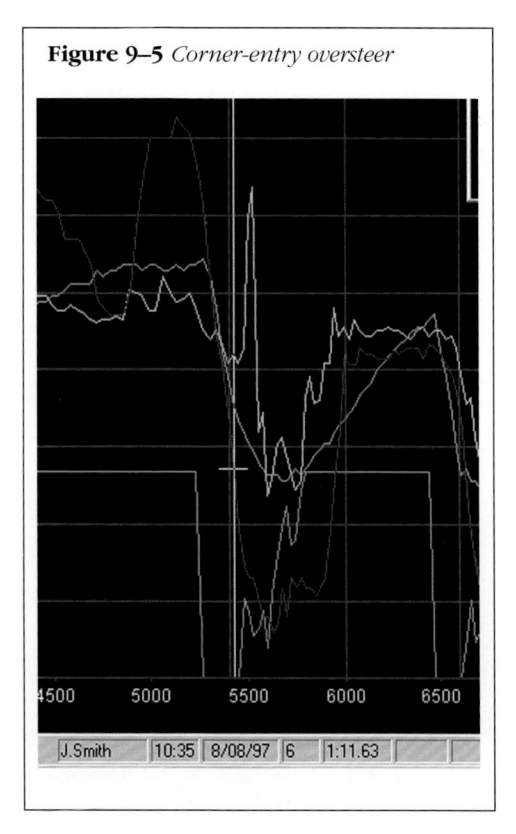

Figure 9–5 *Corner-entry oversteer*

ing is wound back to the neutral position. Lateral G is declining to below the peak value for this corner, showing that the radius of the car's line is now opening out at the corner exit.

Understeer

In Figure 9-6 there is a rare example, from this car in this session, of what we can interpret as corner-entry understeer. As we have seen this car seems to have a fairly tail-happy set-up, exhibiting power-on oversteer on the exit of most of the main corners, and also corner-entry oversteer in the previous example. However, at Coppice Corner on one lap the car exhibited brief understeer. At the point where the cursor is set in Figure 9-6 the steering is just about to be turned to the right to enter the corner. Speed is reducing rapidly and the driver is off the throttle, and therefore probably (although not necessarily) on the brakes. Right steering lock is then applied fairly rapidly and at the same moment the throttle is also squeezed. Steering angle reaches a peak early on, as does lateral G, but notice that the lateral G reaches its peak just after the steering. This could point at mild understeer, with the best grip being achieved just as lock was paid off again, and after the slowest part of the corner.

The occurrence is brief, and the car was rebalanced with the increasingly rapid throttle opening that took place very soon after the slowest part of the corner. Lateral G decreased and steering lock was lessened once the throttle was wide open. The distinction with the corner-entry phase data of the traces in Figure 9-7 from the previous example are clear – here the steering reaches its peak slightly later and the lateral G trace matches the steering trace very closely rather than lagging just behind it. The example in Figure 9-6 corresponds with what we feel from the driving seat with corner-entry understeer – the steering is wound on more than it should be for the radius of the turn at any particular point, and because the car is

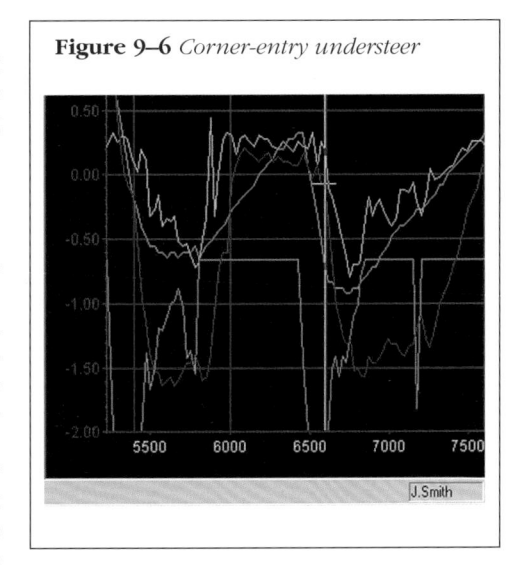

Figure 9–6 *Corner-entry understeer*

not gripping as well as it could, lateral G is less than optimum.

Understeer at any point in a corner will also be evidenced by the steering angle being greater than it should be for the radius of the turn. But how can you tell if this is the case from the data? First, try looking at the actual steering angle in corners of similar speed. Similar speed

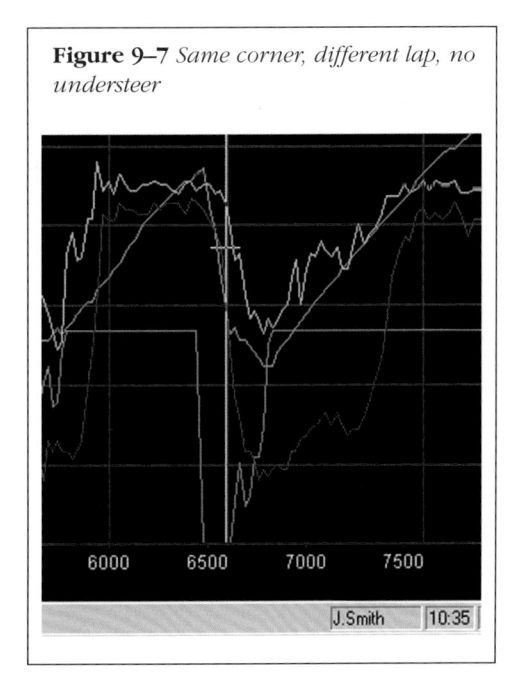

Figure 9–7 *Same corner, different lap, no understeer*

Understeer will show up as excessive steering angle and reduced lateral G in the data. (Tracey Inglis)

corners are likely to be similar radius corners for the reasons stated earlier in this chapter. So if the car exhibits a similar handling balance in corners of that speed then you would expect similar steering angles to occur, all other factors and inputs being equal. If the steering angle is less than it seemingly should be, then the car may well be oversteering, and if it is obviously greater than it should be then the car may well be understeering.

Furthermore, the shape of the steering trace in a given corner should match the shape of that corner. The lateral G trace is usually going to show peak cornering force where the corner is tightest, and the steering should be expected to follow a similar pattern. Any obvious departure from this pattern may be cause for investigation, and may point up handling quirks or alternative lines being tried by the driver (intentionally or otherwise!). Again, look for consistency lap to lap. Inconsistent laps, that is, ones that do not match the usual pattern in a given corner, may reveal the occurrence of a problem or mistake.

If we take a look again at the traces shown in Redgate Corner, but this time by zooming in a bit closer (see Figure 9-8), we can see that the more conservative use of the throttle on this run produced mid-corner understeer here. The giveaways are the early lateral G peak, which then shows a gradual decline

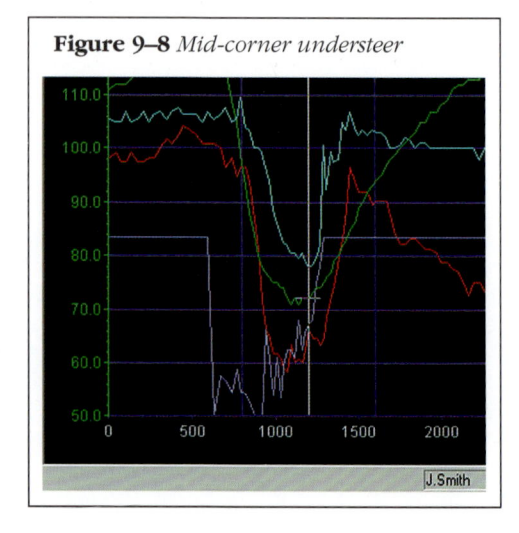

Figure 9-8 *Mid-corner understeer*

through the mid-part of the corner, accompanied by a late steering angle peak. In other words, in the middle part of the corner the cornering force is declining even though more steering lock is being applied. The fix occurs when the driver gets on the throttle sufficiently hard and a steering correction is required. This actually adds another small lateral G peak before the corner opens out and the cornering force tails away.

Figure 9-9 shows these two laps plotted together, the grey traces representing the 'datum' lap that we looked at earlier, and the coloured traces represent what was actually the preceding lap. The difference in the steering traces is very marked, reflecting the equally different throttle technique used. The lateral G traces are actually fairly similar, although the peak on the datum lap was higher. As an indicator of the performance difference, the speed trace (not shown for reasons of clarity) on the datum lap shows more speed was achieved on corner-exit with the understeer-cancelling assertive throttle method. Furthermore the split report shows nine hundredths of a second gained in this one corner through the assertive approach. Once more, the technique worked with this driver, this car and this set-up. That is not to say that this is how a racecar should be driven necessarily – that's not what this book is about. Rather, this is the type of occurrence you can study and derive from data logging! Once you've spotted the problems, the solutions are many and varied, and are down to you!

Problems in interpretation can arise if a car exhibits, or a driver induces understeer on a consistent basis. This is where a mathematical approach might be of help, providing your data logger will let

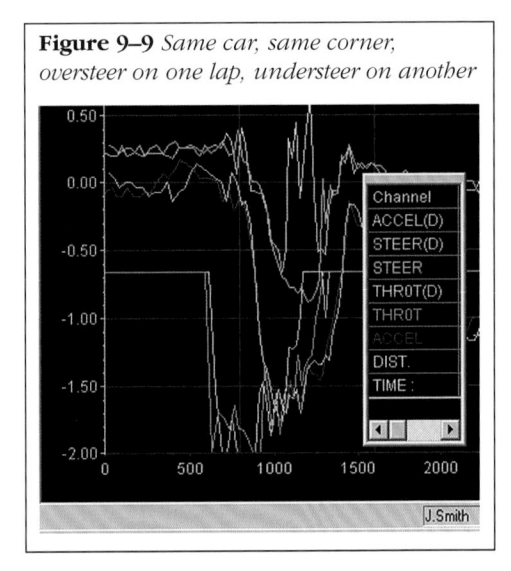

Figure 9-9 *Same car, same corner, oversteer on one lap, understeer on another*

you do this. Even if the DAS software you have does not specifically enable so-called 'maths channels' to be plotted, in all probability you will be able to do some maths with the data by importing it into a software package (a spreadsheet) that does allow you to do maths. More on that in a couple of chapter's time.

Cornering lines

The line the driver takes in a corner can also influence the shape of the lateral G trace. Depending on the precise geography of the track, if he or she turns in early and 'takes an early apex', then the line followed by the car will mean the tightest radius will be encountered late in the corner. This will also give rise to a late lateral G peak, mirrored by a late steering peak if the handling is balanced and neutral. Conversely a late apex will very likely produce the tightest radius early in the corner, which will yield early lateral G and steering peaks, again assuming the handling is balanced.

Chapter 10

Longitudinal G

AT THE RISK of stating the obvious, longitudinal G measures accelerations at right angles to lateral G. The axis of the accelerometer that makes longitudinal G measurements is aligned with the fore and aft axis of the car, and it therefore measures the forward accelerations generated when the car is accelerating under engine power. But it also measures the rearward-directed accelerations generated when the car is slowing down, either through the combination of aerodynamic and frictional drag when the driver just lifts off the throttle, or when the brakes are applied.

As with the lateral forces and accelerations arising from cornering, it is sometimes confusing to think of slowing or braking as being directed rearwards, but if you think about the direction of the force that must be applied to slow a car, it has to oppose the direction of travel. If the car is travelling forwards, then to slow it down you have to push backwards. Thus, decelerative forces are directed rearwards, and so the deceleration arising from this force is also in a rearward direction. Conventionally, forward accelerations are regarded as positive (+) and rearward accelerations as negative (–).

Once again, longitudinal G is usually stated in units of G for convenience, and like lateral G it too is dependent on the coefficient of friction or inherent grip offered by the tyres, and the vertical force acting upon them. If significant aerodynamic downforce adds to the combination of static and dynamic 'mechanical' vertical forces, this increases the longitudinal G that can be generated (at the speeds where the downforce becomes significant that is). Tyres can of course only generate so much force in any direction, so if maximum lateral G is being generated then the tyres will have little or no remaining capacity for generating longitudinal G at the same time, and vice versa. We'll look at this concept from the data logging viewpoint in more detail later in the chapter.

The maximum longitudinal G that a given car can generate can perhaps be roughly estimated from knowledge of the coefficient of friction of the tyres. But whereas maximum cornering force is the sum of the contributions of (up to) all four tyres, maximum acceleration is only ever going to be the sum of the contribution of the driven tyres, which in most cases is just two at the front or two at the rear. Of course, four-wheel drive is an exception to this, but in general this means that absolute maximum acceleration G, even in the lowest gear, will be considerably less than maximum lateral G.

Maximum braking forces, although they are generated by more than just a pair of tyres, are also unlikely to reach the levels of peak lateral Gs in most cases although they should not be too far off. Maximum potential can be found out practically with simple straight line brake testing,

Maximum acceleration is limited by the contribution of the driven tyres.

All four tyres contribute towards braking.

once tyres, brakes and everything else are up to working temperature. Exceeding the maximum grip that the tyres can generate longitudinally results either in wheelspin under power, or wheel lock-up under braking. Other factors such as circuit geography and surface can affect the maximum potential longitudinal Gs in acceleration and braking. Clearly, braking uphill is going to enable greater G values, as is accelerating downhill.

Let's take a look at our Donington Park data again (see Figure 10-1), and specifically at the longitudinal G trace (purple) along with the speed trace (green) generated by the Formula Vauxhall Lotus. Again, Pi Research's name for longitudinal G is 'InLin', short for in-line G. Elsewhere you'll find abbreviations like G Long, LONGAC and LnG.

The first thing that's obvious on the long G (as perhaps we can abbreviate it) trace is that there is a series of dips, showing negative G, that coincide with the speed drops prior to each of the main corners. This is where the driver brakes to cut his speed for corner entry. The second thing that is apparent is that these dips are not all to the same level. Corners 1, 3 and 4 produce peak values of around −1.2 to −1.3G, whereas corner 2 produces just −1.07G. Braking for the chicane (corner 5) however produces a peak of −1.50G. We'll come back to this feature in a while.

Another feature that is of interest is the width of the dips, because this indicates, in effect, how long the car spent slowing down, or in other words, how long the driver was on the brakes. This can be a useful point to look at when comparing data from lap to lap, or with other drivers. It is also possible to compare at what point the car began to slow down, and where and when it began to accelerate again, each useful comparisons when examining driver technique.

The shape of the long G dips is also important because it tells you how the driver was using the brakes. The dips on this trace are narrow-based, which is said

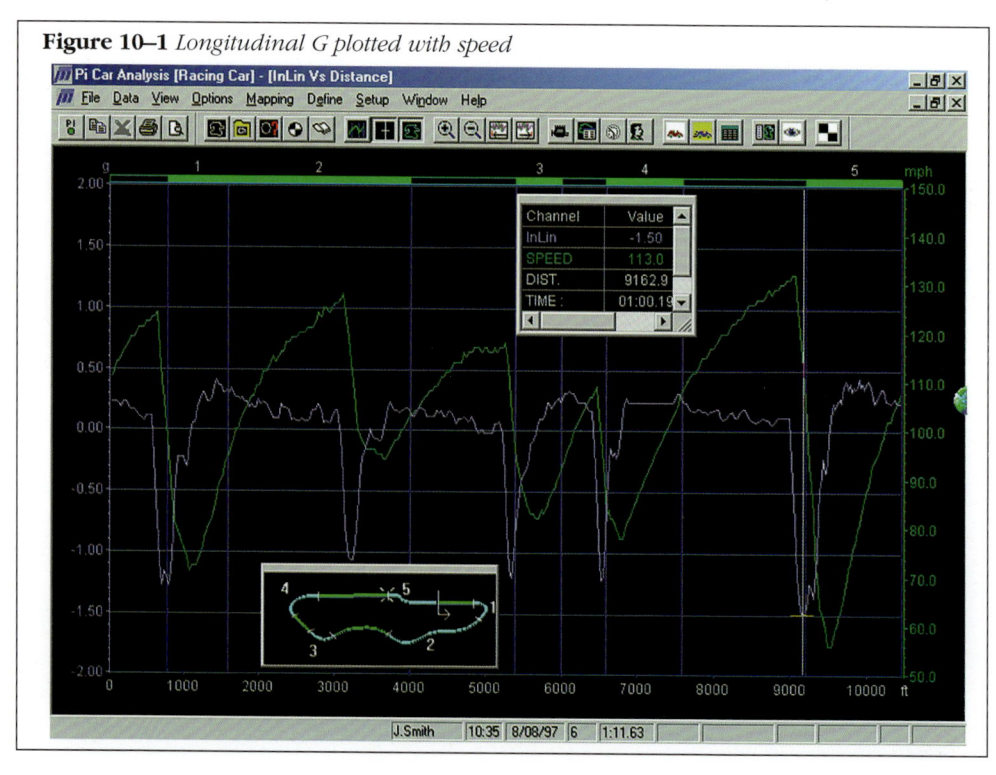

Figure 10–1 *Longitudinal G plotted with speed*

to be typical of a 'wing' car, that is, one which generates significant downforce. A non-wing car, one which does not create much, if any, downforce, would probably generate broader troughs. We'll get into the whys and wherefores of that a bit later too.

The dips due to braking are immediately obvious. But the long G trace also shows where, and by how much the car was accelerating. As an example, from about 1,100ft, where the slowest speed in Redgate Corner, the first corner, was reached, the car accelerates. Long G climbs from zero to 0.41G at about 1,450ft. Notice then how the acceleration tails off. This is due to two things; first, the driver changes up two gears in this section out of Redgate (as we saw earlier on an rpm trace), and acceleration is less in higher gears; and second, as speed builds, so does aerodynamic drag, and this is fundamentally what causes the speed to slowly level off as it approaches its highest values on any track. Long G is in this instance a measure of this levelling off in speed gain, so long G also tails off at speed. Eventually, as a car approached its actual maximum speed, as limited by the total of frictional and aerodynamic drag, long G would tend towards zero because all available power would be expended overcoming drag; there would be no power left for acceleration.

That's probably dealt with the generalities of long G. Let's delve a bit deeper once more to see what can be learned from the detail. Figure 10-2 shows the same lap, but once more we've zoomed in on Redgate Corner. The lateral G trace (red) has also been displayed to help figure out some of what's happening. First, the white cursor line has been positioned at the point where long G (the upper purple trace) reaches its first maximum of 1.28G for this braking zone. Notice that the long G dip has built rapidly, showing that the driver was on to the brakes rapidly and sharply.

Immediately to the right of the cursor line though is a small upward blip in both the long G and throttle traces. The blip in the throttle trace is where the driver actually did blip the throttle on a heel-and-toe downshift, but the small spike in the long G dip demonstrates that braking power was briefly reduced (to −1.16G) during the throttle blip, the driver having reduced pressure on the brake pedal slightly during the throttle blip. After this, long G again dips to −1.28G as full brake pressure resumed.

Over the next 0.6 seconds the long G tails away to −0.55G as the driver eases off the brake and then moves his foot to the throttle, and at the same time the lateral G starts to climb as the car enters the corner. This to my mind (speaking as an amateur racecar driver) is one of the toughest things to master in a competition car – what is often called 'trail braking'. This is the technique of maximising the grip that your tyres offer by balancing braking and cornering demands whilst entering a corner, and it involves gradually easing off the brakes during the turn-in phase, as distinct from the rather more simplistic 'do all your braking in a straight line before you turn in' method.

As mentioned earlier, your tyres offer only so much grip in any one direction,

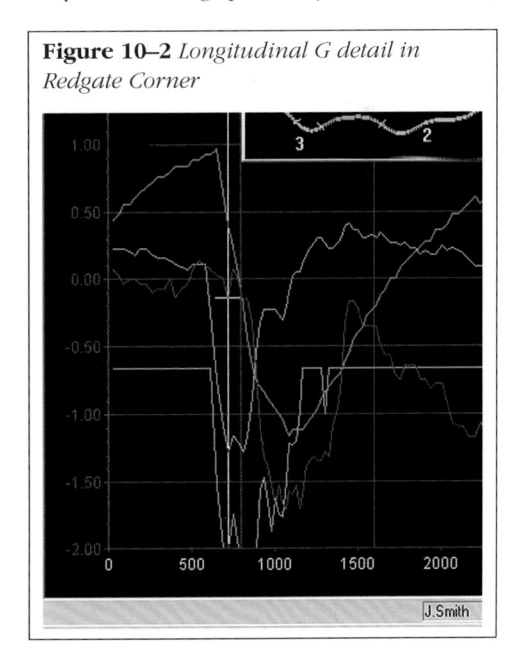

Figure 10–2 *Longitudinal G detail in Redgate Corner*

Figure 10–3 *The 'traction circle'*

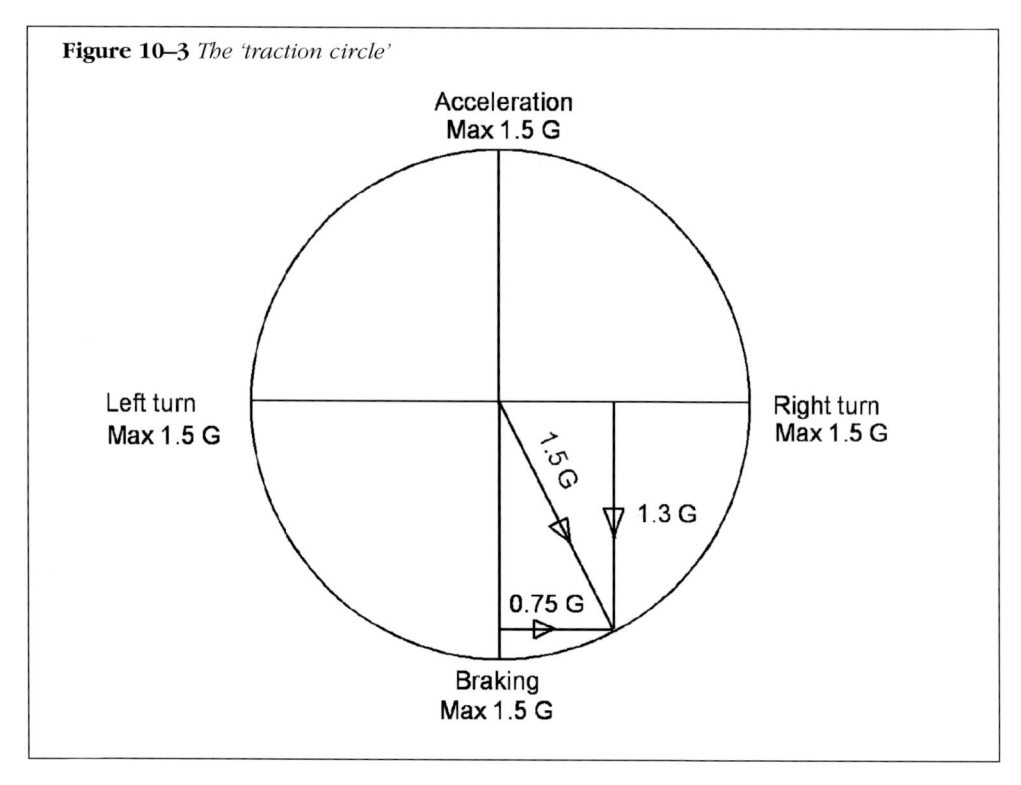

- Acceleration Max 1.5 G
- Left turn Max 1.5 G
- Right turn Max 1.5 G
- Braking Max 1.5 G
- 1.5 G
- 1.3 G
- 0.75 G

but they don't really care which direction the total force adds up to, so long as it doesn't exceed its capacity to grip the road. So whilst maximum braking force is being generated, little or no cornering force will be tolerated by the tyres. Equally, while maximum lateral grip is being generated, little or no braking or accelerating will be tolerated. In between those extremes though, the tyres can deal with an amount of combined braking and cornering force, or combined cornering and accelerative force.

This gives rise to the concept of the 'traction circle', which helps to explain that tyres offer a maximum amount of grip which can be a combination of lateral and longitudinal demands. In Figure 10-3 an imaginary traction circle is drawn for a tyre that can generate a maximum of 1.5G in any one direction. From this you can see that, in the example drawn, if the tyre is called upon to develop –1.3G longitudinally at some instant in the corner-entry phase, it can simultaneously generate (+ or –, left or right) 0.75G in lateral G. If the longitudinal and lateral forces were exactly balanced, such as might happen a little further into the imaginary corner, the tyre could generate 1.06G both in braking and cornering force.

This is not an easy concept to get your head around. It might seem as though generating 1.06G in both directions ought to add up to 2.12G total when we said that the tyre can only generate 1.50G total. But because the longitudinal and lateral forces are at right angles to each other, they don't just add up together. In fact, for the mathematically inclined, the way to calculate the maximum combined capacity of the tyre is as follows:

total G^2 = [(lateral G)^2 + (longitudinal G)^2]

Enthusiasts of mathematics will immediately recognise this as Pythagoras's theorem relating to the sums of the squares on a right-angled triangle. At last, you now have a practical use for it! For non-enthusiasts of maths, and there are a lot of us I'm sure, let's use a calculator and not concern ourselves that an Ancient Greek, however clever and well intended, may have invented the formula in the first place. It's still a useful formula!

So back to the data again. We had noted prior to the above mathematical diversion that the long G had tailed away to −0.55G, which still represents deceleration, but notice that the speed trace shows this lower rate of deceleration too. In fact the driver is just beginning to open the throttle, so we can be fairly sure that he has now come off the brake. The small throttle openings are now to balance the drag (aero and friction) of the car to sustain some speed towards the corner apex, although in truth the car is still slowing here.

Let's stop briefly and look at the total G being developed here. Where the G traces cross, which happens to correspond to where the throttle opens, lateral G is −0.76G and long G is −0.55G as we said above. Finding the square root of the sum of the squares of these values gives:

$$\sqrt{[(-0.76)^2 + (-0.55)^2]} = 0.94G$$

Note: the negative signs magically and conveniently disappear when you square the values.

So at this particular instant the car is developing a combined total of 0.94G. We know by examining the braking G peaks that −1.2 to −1.3G was developed before most corners, and that typically −1.6 to −1.7G peak was generated in most corners, so it looks as though the entry to this corner is not fully exploiting the available grip of the tyres at this point if a combined G value of just 0.94G was being generated. It would be possible to make the calculation at various other points in the corner entry phase too, and build a picture of exactly where the driver is losing out.

As I said earlier though, I feel this is the hardest job for a competition car driver to do, so far be it from me to criticise! But the data does let you analyse what is happening. In the next chapter we'll look at ways of using the DAS software, and other computational methods, to get a clearer picture of this scenario more rapidly. It will become apparent that even drivers in categories capable of generating 3G to 4G in cornering and braking (and there aren't too many of those . . .) do not always use all the grip that is available to them in this situation, something from which the rest of us can take some comfort!

Back to the data again. By the time the car reaches its minimum speed in this corner, the lateral G peaks at −1.71G but the long G has now tailed off to −0.21G, still just slowing down. Then, once past the slowest point, clearly the speed starts to rise again, and sure enough the long G goes positive. Lateral G actually reaches a second peak of −1.71G, and then this starts to decline in value as the car now exits the corner and the driver is hard on the throttle.

At the first small peak in the long G trace, 0.30G is being generated in acceleration, and −1.35G in lateral G is being simultaneously generated. The square root of the sum of the squares of these values is 1.38G. This is well below what the car should be capable of tolerating before losing grip, and yet we have seen that it went into oversteer in this corner, to the extent that the driver had to make steering and throttle corrections. This further emphasises the point that by overloading the rear of the car with an aggressive throttle technique, grip as well as balance was lost. And yet, recall that this particular corner was done (seemingly) very quickly on this lap . . .

It was mentioned earlier in this chapter that uphill braking zones can enable higher long G to be generated. Equally, downhill braking zones should be expected to yield lower braking G, and this explains the lower long G dip prior to Old Hairpin, or corner 2, on this track (see Figure 10-4). The approach is via the fast Craner Curves, a downhill section of track. It can be seen that there is little distance between the lateral G peak of the left-handed Craner Curve, where the cursor line is positioned, and the point where the brakes are first applied, as shown by the long G trace and the throttle trace. The long G reaches a peak value of −1.07 here, quite a bit less than in the other braking zones on the track.

This will in part be due to the downhill approach, but also to the fact that less speed has to be disposed of for this corner than the others. This leads to reduced braking force, even though one might think that the same deceleration would be applied for less time in such a case. The reality is that less aggressive braking is often better because it destabilises the car less, and causes less change of pitch, or nose-down angle, more critical perhaps on a car that generates a large proportion of its downforce from its underbody, but still important on any 'wing car' if the front wing is not to be brought so close to the ground that it can upset the aerodynamic balance of the car.

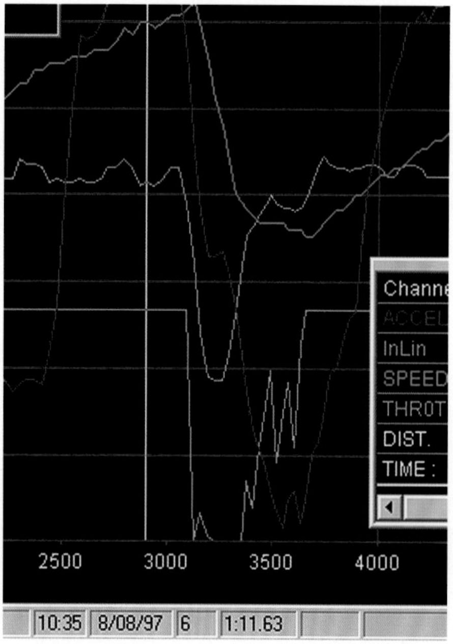

It was also mentioned earlier that wing cars should generate sharp longitudinal G dips, as shown in the example data here. Why? Because at high speed the downforce being generated is providing greater grip than is available as speed decreases. Thus, maximum braking force is available as soon as the brakes are applied, but the maximum braking force available declines as the speed, and hence the downforce reduces. Hence, after a rapid initial build up of negative G, the long G trace should then show a decline as the driver eases pressure off the brake pedal.

A non-wing car, that is, one that generates little or no downforce, will have fairly consistent grip available at any speed, and hence the maximum long G value will be independent of speed. This will generate smaller long G values than are possible with a downforce-assisted car (assuming tyre grip is no greater), but also longer duration long G peaks, or dips if you prefer, because the driver will have no cause to back off the brake pedal pressure as speed reduces. The rate at which long G builds will also be somewhat slower because maximum grip only arrives once the forward weight transfer initiated by braking has occurred. So in general terms the long G troughs will be wider and shallower with a non-wing car.

As with the other data channels we've examined along the way, driver (or lap) comparisons can be extremely useful. Longitudinal G traces enable at-a-glance analysis of driver braking technique. Take a look at Figure 10-5, which shows a pair of long G and speed traces of two laps around Donington overlaid, actually done by the same driver during the same session. The green speed trace is the same lap as the purple long G trace, and the orange speed trace ties up with the yellow long G trace. The differences in long G, and the effect they have on speed, are readily apparent.

Figure 10–4 *Braking downhill generates lower longitudinal G*

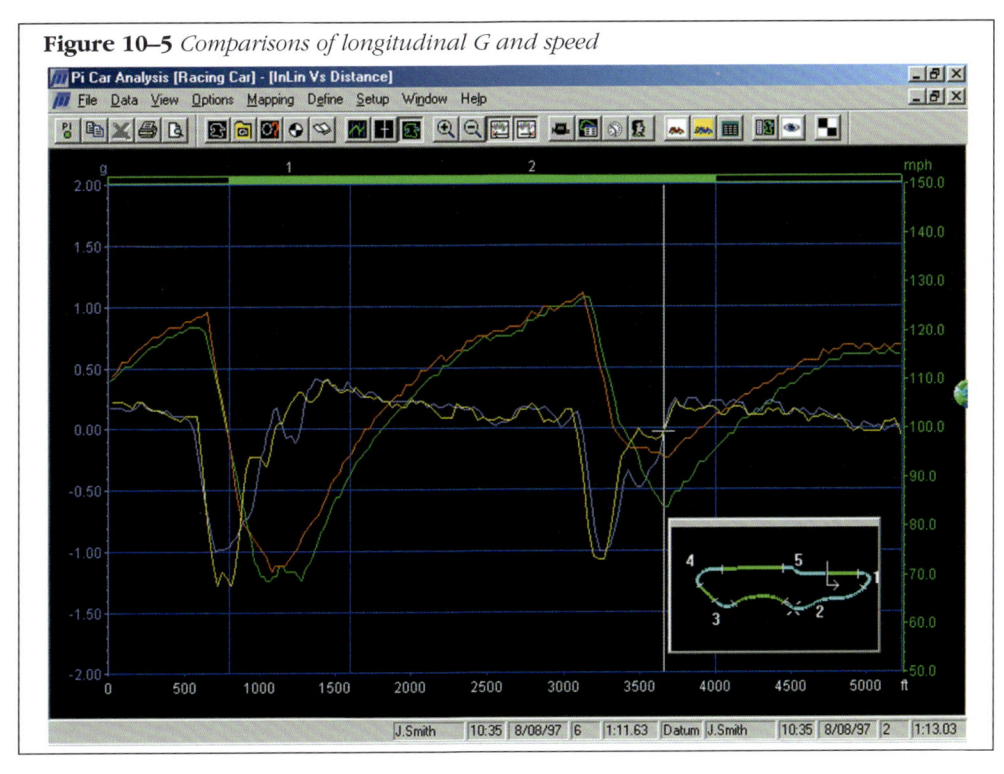

Figure 10-5 *Comparisons of longitudinal G and speed*

Look first at the differences in the long G traces during the braking phase for Redgate Corner, which commences at around the 500 to 600ft distance into the lap. A number of things are apparent; first, the purple trace starts to dip earlier than the yellow one; second, the purple trace reaches a smaller peak G value; third, the purple long G dip is broader. All of this means that on the 'purple' lap the driver started braking sooner, he braked less hard, and he spent longer on the brakes.

Looking at the speed traces now, we see that on the 'green' lap the car approached the corner at a slower speed, roughly 2mph down compared to the orange lap, and speed reduced earlier and initially more gently than on the orange lap, which ties in with the earlier and more gentle braking evidenced by the purple long G trace. The traces then more or less overlap for about 150ft, at which point the green lap is slightly faster for about 100ft. This coincides with the slightly earlier release of the brake as shown by the purple long G trace.

It is at the latter part of the corner-entry phase that the speed traces show the biggest differences though. From about 930ft to the minimum speed point, the orange speed trace is faster than the green trace. This corresponds with the more rapid release of the brakes on this lap, shown by the steep rise in the yellow long G trace, but is more due to the throttle (traces not shown here) being opened sooner to drive the car into the slowest point in the corner.

A quick look at the braking for the next corner shows a somewhat different pattern. Initially the driver did a better job on the lap shown by the green speed trace and the purple long G trace, braking slightly later. This saw more speed carried into the corner, as shown by the green speed trace. But peak long G was slightly less on this lap, and the long Gs (that is, the brakes) were released more slowly, which saw the orange speed trace begin to catch up the green one. The marked difference in minimum corner speeds a bit later in this corner was mainly down to the timing and assertiveness of throttle application driving into the corner apex and nothing really to do with braking, except for the brake being released a little bit sooner on the orange speed/yellow long G lap.

So it is clear that studying the longitudinal G trace yields very useful information about the timing and technique of brake application, and the impact on speed carried towards the corner apex. Using the throttle trace too helps to define the timing of braking more clearly. Acceleration can also be examined, and looking at the combination of longitudinal and lateral G on corner-entry and corner-exit can indicate if a driver is making the most of the total grip available.

By studying the data from the six channels we have looked at over the last six chapters, it is apparent that we can get a pretty good idea of engine, chassis and driver performance. By learning the various patterns and shapes that occur in the data traces, it becomes possible after some practice to 'eye-ball' the graphs and quickly get an idea not only of where on a given track the data corresponds to, but also whether the shapes of the traces indicate problems or successes that can be learned from.

In truth you can get a surprisingly long way to knowing what's going on by looking carefully at just rpm and speed. But with the help of throttle, steering and the two G channels things become that much clearer, and confidence grows in diagnoses of problems or other events.

Chapter 11

Software extras

EARLIER IN THE book mention was made of other types of graph plots and information analysis that are made possible by the analysis software that comes with your data acquisition system. What's available to you is, by and large, related to how sophisticated your DAS is (that is, how much you paid for it, in effect).

X-Y plots
The most obviously useful of the additional graph plots that's available is the X-Y plot, or scatter plot as it is also known. The X-Y plot allows you to make a graphical comparison of two different sets of data that may or may not be related. If you already know they are related, you might want to search the data for the basic 'shape' of the relationship, or study it more closely still to look for anomalies. If you are as yet uncertain as to whether two sets of data are related, then plotting them as an X-Y plot will give you a much clearer idea.

So what is the anatomy of an X-Y plot? It's a type of graph which, instead of having time or distance along the horizontal, x-axis as we've become accustomed to throughout this book, has one of the other logged parameters along the x-axis, and another logged parameter on the vertical, y-axis. Let's take a look at the example of an X-Y plot in Figure 11-1. This is a plot generated by a Magneti Marelli DAS fitted to a Formula Renault Sport single-seater racecar. As you can see the chart dates back to 1997 and comes from Croft circuit in north-eastern England.

The graph shows speed versus rpm, a familiar-looking plot if you already work with gear charts. Speed is plotted along the x-axis, from 50 to 250km/h, whilst rpm is plotted on the y-axis, from 4,000 to 7,500rpm. The basic shapes formed by the data look fundamentally similar to a gear chart, which one usually sees drawn with nice, clean, straight lines. The X-Y plot here is made up of the data collected over a single lap, and each data point is plotted as just that, a point, on the graph.

You can tell a number of simple, basic facts from the plot; for example, four gears were used, and going by the number of points plotted in the highest three ratios compared to the lowest ratio, most of the lap was done in the top three gears. Next, you can see that maximum rpm in each gear was around the 7,000rpm mark, with a few points exceeding the 7,150rpm line; the rev peaks in the two intermediate gears were virtually identical, and so on.

But other interesting facts emerge too. The lowest gear might only have been used for a short duration, but the next lowest gear was used at speeds well within the range of the lowest gear, allowing minimum revs to drop as low as 4,500rpm. This perhaps raises a question – was the driver using his engine's power band to best advantage? Maybe a judgement had been made that less time was lost by pulling from low minimum rpm

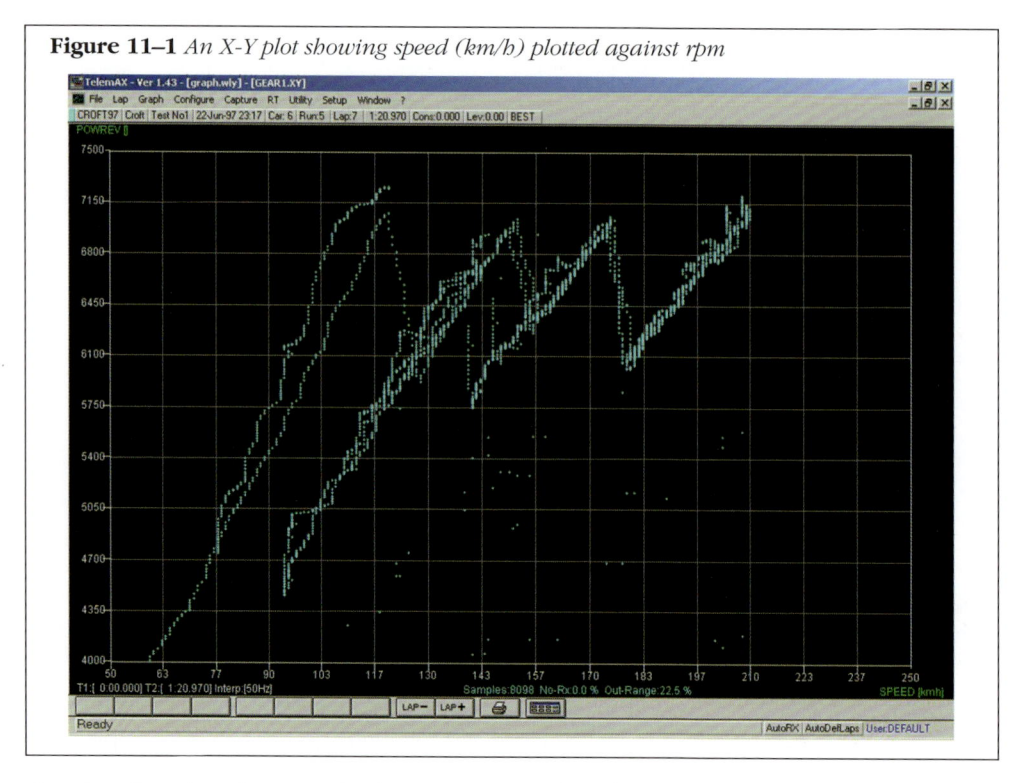

Figure 11-1 *An X-Y plot showing speed (km/h) plotted against rpm*

than by changing down to the lower gear and then up again. Or maybe the race engineer could look at this data and then say to the driver: 'Why are you asking the car to pull from these low rpm?' The same pattern can be seen at the overlap of the two intermediate gears – the revs occasionally fall into the power band of the lower gear. But of course, what we cannot tell from this plot is whether this was just on the overrun, that is, during deceleration rather than acceleration.

We can see that the rev drop into top gear was actually greater than the rev drop between the two intermediate gears. This seems to be at odds with the theoretical best solution for staying in the most effective part of the power band at higher speeds. We can also see that there are a lot of data points in the lower intermediate gear, indicating more time was spent in this gear than the others. These two facts suggest that the lower intermediate gear ratio was picked as a fairly tall ratio to match the corners on the course rather than just fit with a generalised idealistic pattern.

You may be wondering why there is as much scatter in the data points when simple maths tells you that a given rpm value should equate to a given speed value, and the data should therefore produce the same straight lines as on your gear chart. There are a couple of main reasons. In the lower gears especially, torque at the driven wheels can easily exceed available grip, which leads to wheelspin, which in turn leads to anomalously high rpm values for the speed recorded at the same instant. Secondly, with just one wheel speed sensor, say on the left front wheel for a right-handed circuit, the path followed by the left front wheel will mean it tracks a longer distance around right-handed corners than the inner wheels, and vice-versa on left-handed corners. The rpm (with no wheelspin) will equate to whatever engine revs, in effect, are dictated by the rear wheels and the differential, and so differences can arise between expected rpm and logged speed. Obviously, wheel lock-up under braking of the wheel with the speed sensor can cause brief anomalies too.

So, not only can we see the value of X-Y plots, but also it is clear that speed versus rpm can be a very useful use of X-Y plotting.

In the previous chapter the concept of the traction circle was discussed. This involved the mathematical treatment of the lateral and longitudinal G values occurring at any one instant to assess if the driver was making the best use of the total grip available at all times, especially on corner-entry and corner-exit. X-Y plots of lateral G plotted against long G are a more convenient way of assessing this.

Take a look at Figure 11-2. Despite appearing like an intergalactic star map (too much *Star Trek* viewing again . . .) or a failed effort to paint with a blocked spray gun nozzle, this graph actually shows lateral G (x-axis) plotted against longitudinal G (y-axis) over the course of a number of laps in a single outing. You can see from the G values ranging up to −4G in braking (long G), and 2.5 to −4G in cornering (lateral G) that this data must have come from a sophisticated racecar, because there are not many competition categories that enable the generation of such large forces.

Let's first look at the fundamental anatomy of the chart again. At first there just seems to be a general scatter of points creating a rough triangular or heart-shape. But look at the graph with half-shut eyes and denser regions where data points are more concentrated begin to emerge. At the intersection of the two zero lines there is a dense patch of points which obviously corresponds to the uneventful parts of the course where no braking or cornering forces were being generated. However, this central patch does extend to longitudinal G values of +1G to −1G, covering the range of straightline acceleration under power

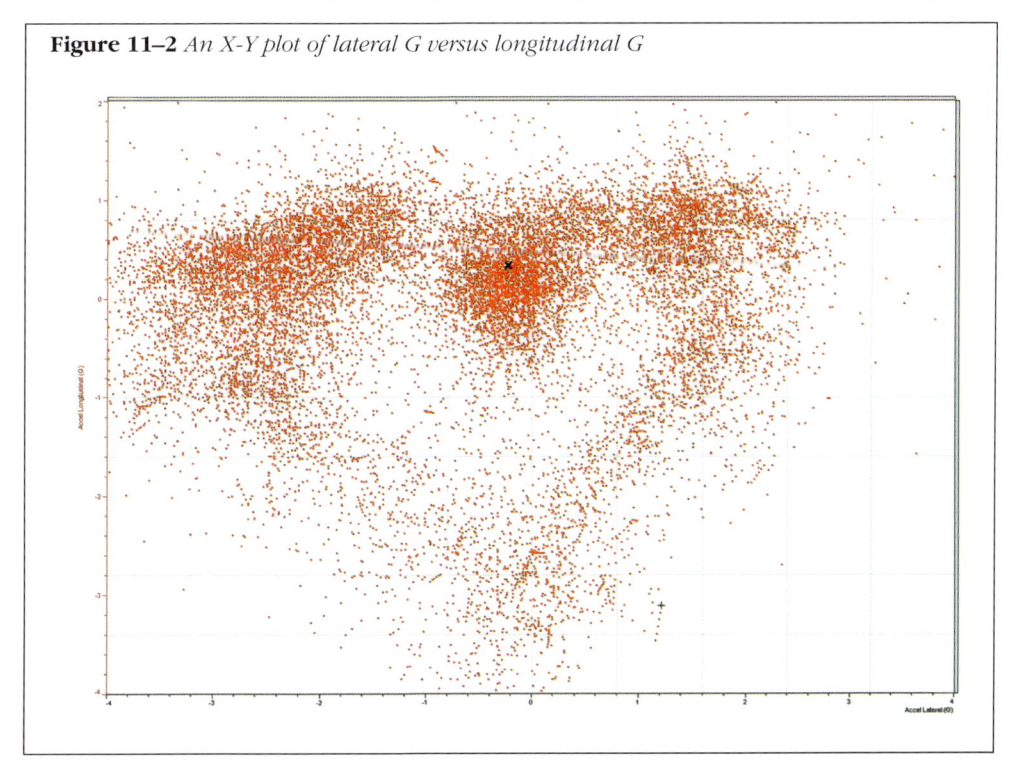

Figure 11-2 *An X-Y plot of lateral G versus longitudinal G*

(zero to +1G) to modest deceleration, possibly when just lifting off the throttle (zero to –1G).

To either side of the central patch are dense regions where high lateral G values were recorded. Maximum lateral G values occur, not surprisingly, where zero longitudinal G was being registered (and vice versa). However, high lateral G values were also logged when both positive and negative long G values were occurring, which creates the curved outer extremities of the denser scatter of points. Similarly there is a (less dense) scattering of points stretching from the peak lateral G values down to the peak long G values, and this creates the other sides of the heart shape.

So what happened to the traction circle? The heart shape here is evidently not circular. In fact, in common with most X-Y plots of lateral versus longitudinal G, the majority of the points on this graph occur along more or less straight bands joining the lateral and longitudinal maxima, with a patch around the 'non-event' zero region.

As an amateur competition car driver I find this very comforting! Why? Well we deduced above that the competition category from which this data comes was a senior level one, going by the high G values. From that we can also conclude that the driver was no slouch either because only drivers in the 'very good to excellent' bracket get to drive such racecars, whatever anyone serving out sour grapes might try to tell you!

And yet this data shows that even this driver doesn't seem to have been extracting all the available grip from the tyres . . . and if a driver at this level can't dance on the limit the whole time, then that makes me feel a whole lot better! In truth, I do not know the exact circumstances that led to the creation of this particular sample data chart – but I still find it comforting!

But, if you'll forgive the pun, back to the plot. The fact is, the traction circle is never going to be circular, especially with a car that generates significant downforce as this one must have done in order to have recorded such high cornering and braking G values. Maximum acceleration will be limited not only by available grip but also by available power and torque. We know that maximum acceleration occurs in the lower gears in which downforce is much less significant (limited grip is therefore available), and even with 800+ horsepower and traction control, such as a Formula 1 car might have at its

It's comforting to know that even drivers in top categories struggle to fully exploit the 'traction circle' . . .

disposal, tyre limitations and two-wheel drive put a cap on maximum longitudinal G under acceleration. So the top half of the traction circle will always be flattened off with this type of car.

As to the lower half of the circle, this will only be circular if the car and driver can actually stay on the limit of available grip whenever entering, negotiating or exiting a corner, something I still maintain is very difficult to do (maybe that's why I've remained an amateur driver). If you draw an imaginary (or real) half-circle on Figure 11-2 centred on the zero intersections, with a radius that reaches out to 4G, you can see very clearly that the data points are close to the circle's limits when either maximum lateral or long G is being generated, but not when both lateral and long G are being generated. But even a top professional pilot would probably only create anything resembling a circle in the G data on an 11/10ths qualifying run. Performing right on the limit the whole time is something very few drivers in the world can do, and even they occasionally make mistakes.

Nevertheless, and putting aside the negative justifications for failing to reach the limit of available grip, the X-Y plot of lateral versus long G makes for interesting study, and can be used to help drivers try to adapt their technique to extract more, if not ultimate, performance from their cars. Figure 11-3 shows a variation on the theme of X-Y G-plots, this time with longitudinal G plotted on the x-axis and lateral G on the y-axis. The data points are also joined by lines, and in the Magneti Marelli software package from which this sample was taken it is possible to link the DAS-generated track map to spot the location on the circuit of each data point. This enables anomalies to be put into context more readily.

But the traction heart, as perhaps we should now call it, is still evident, albeit now lying on its side, and the basic pattern of the data is similar to the previous example. The data was from the afore-

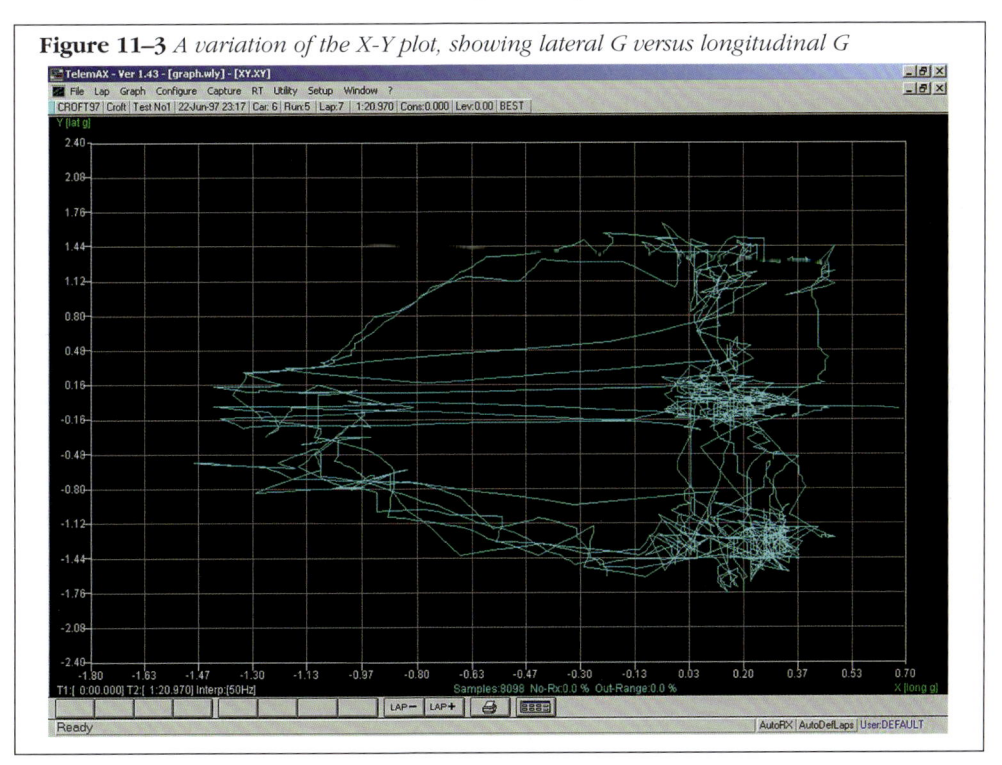

Figure 11–3 *A variation of the X-Y plot, showing lateral G versus longitudinal G*

mentioned Formula Renault Sport single seater at Croft, and the G-maxima are obviously quite a lot lower than in the previous G-plot, ranging from about −1.3G to −1.4G under braking, about −1.6G to −1.7G in the right turns (it's a right-handed circuit), 1.4G to 1.5G in the left turns and about 0.25G to 0.4G under acceleration. Although aided by slick tyres and wings, this car had less aerodynamically assisted grip than the car in the previous example, and a lot less power too (about 180bhp). Yet the traction heart is still there, and a few quick calculations using the sum of squares theorem explained in the last chapter, seems to show the driver is making pretty good use of his total available grip.

Also evident from this plot is that the lateral G only starts to reduce when in excess of around + or −0.3G longitudinal is being generated. This is shown by the horizontal upper and lower regions in the heart shape in Figure 11-3. It is also evident that peak longitudinal G under acceleration (0.30 to 0.35G typically) can be generated at the same time as is significant lateral G. And by the same token, the 'base' of the heart, that is the left-hand end on this plot, shows that high braking G was pulled whilst some lateral G was being generated, although the nature of the lateral G scale and the short-lived long G peaks make it hard to see what lateral G values were being tolerated at the same time as hard braking. In all likelihood, and based on the above, around + or −0.30G lateral would again be a reasonable assumption.

It can be seen that at one point the car pulled about −1.47G under braking (the maximum in the plot) while simultaneously recording about −0.50G lateral. Applying the sum of the squares to these values shows that the driver was pulling 1.55G total, not far short of his peak lateral G values, and showing he was running right up to the limit of the grip available to him at this point. However, that the point stands out on the plot suggests it was an out-of-the-ordinary occurrence when perhaps the driver may have strayed beyond his normal 'performance envelope' because in general the braking G values were lower than in this case, and without anything like as much lateral G occurring at the same time. But you can't help wondering whether he should have been able to reach those kind of combined values all the time . . .

Another often-valuable X-Y plot is one such as shown in Figure 11-4, with lateral G this time plotted against steering angle. Just as a neatly drawn gear chart was analogous to the X-Y plot of rpm versus speed, so the graph you might be lucky enough to get from your tyre supplier showing tyre slip angle versus cornering force is analogous to the X-Y plot of lateral G versus steering angle. In this instance the axes have swapped around, but the information shown is basically similar, with the important distinction that the logged data is from the real world, as opposed to the laboratory. That's not to denigrate laboratory data in any way, because the kind of information that comes with tyres to some favoured competition categories is invaluable. But the logged data is, in its way, more valuable for the ability to see anomalies as well as normal patterns.

Figure 11-5 is a tyre company-produced graph showing slip angle versus cornering force, actually for the 1999 specification F3000 front tyre produced by Avon. Clearly, the cornering force that can be generated rises with increasing slip angle, but at a rate that tails off. Thus, at the higher corning force end of the scale, slip angle may increase a lot yet the cornering capacity does not. Eventually of course, and not shown in this graph, the cornering force actually declines with increasing slip angle, and the tyre loses grip. It's obviously the driver's job to ride close to the peak cornering force as much as possible.

Back to Figure 11-4 then, and you can see that the basic shape of data plotted here, once more from our Formula Renault Sport at Croft, is of the same generic pattern as the tyre company's

SOFTWARE EXTRAS 117

Figure 11–4 *An X-Y plot of lateral G versus steering angle*

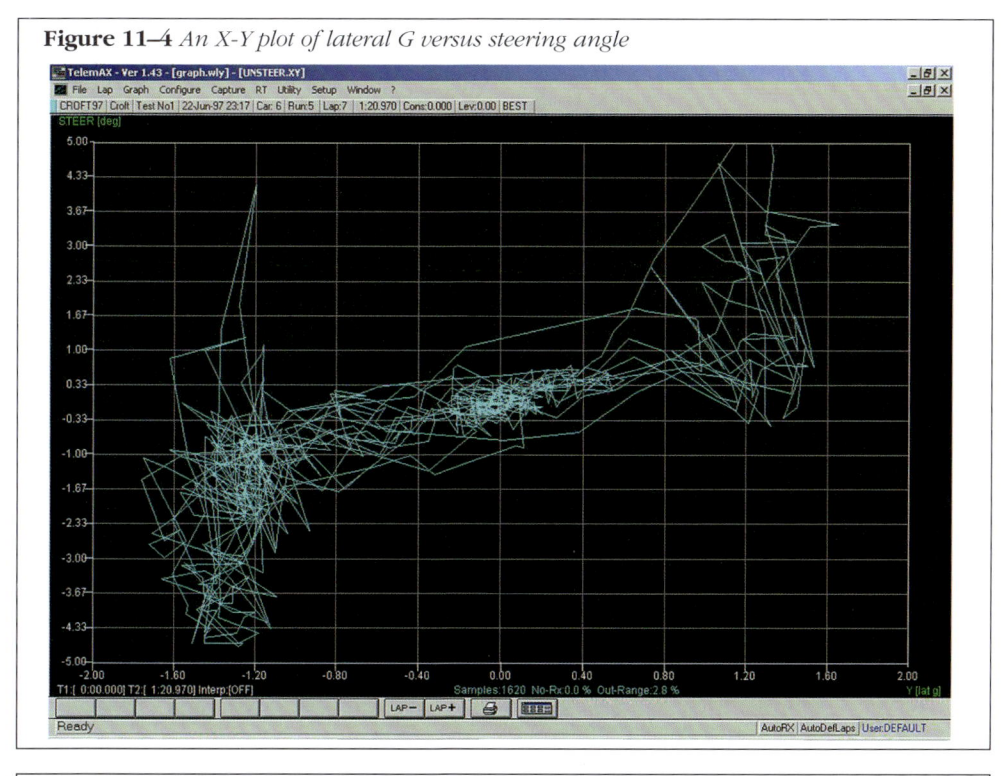

Figure 11–5 *Tyre slip angle versus cornering force at three vertical loads*
(Courtesy: Avon Racing Tyres)

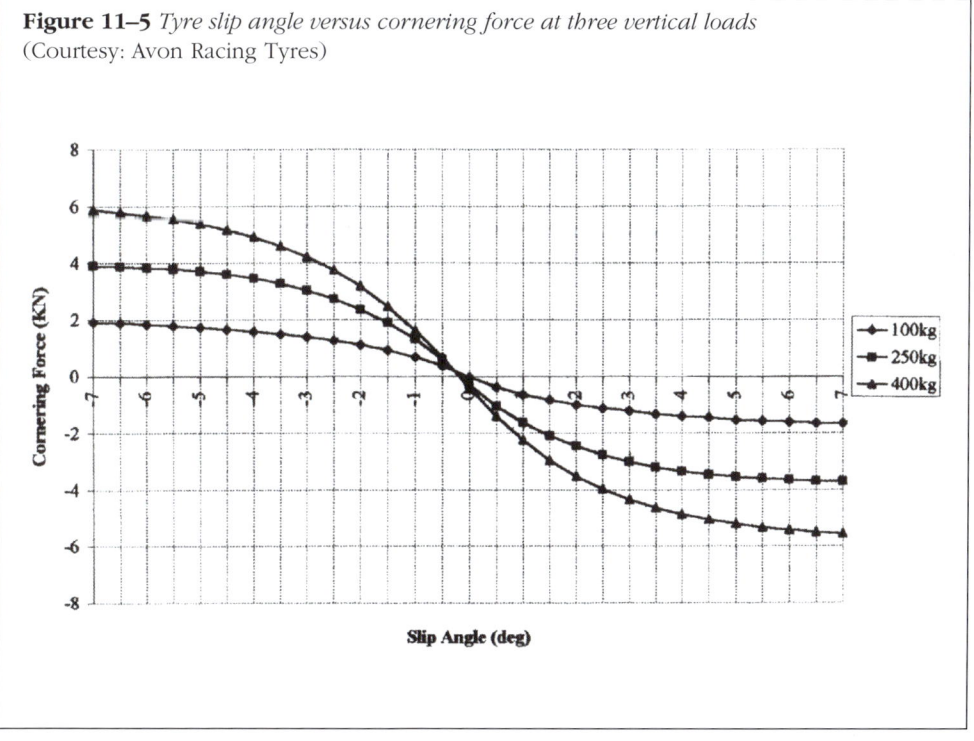

graph. Cornering force (lateral G) rises initially with steering angle, but then even with additional steering angle the cornering force does not increase very much. This is apparent at both ends of this band of data, but is clearer at the left-hand end of the graph where the data represents the dominant (numerically since it is a right-handed track, and in terms of cornering force generated) right turns of this track.

So what else can be determined from this graph? The occurrence of understeer and oversteer. Take a look at the denser cluster of data at the left-hand end of the graph, around the −1.2G to −1.4G and 1 to 2 degrees steering angle region. Now, there is bound to be some scatter in data of this kind, especially arising in the steering trace as the result of feedback from bumps and the input of minor corrections. But basically, you could 'eyeball' this cluster of data and get a feel for the 'normal' steering angle that you expect to generate, say −1.4G, and the usual range of angles that might be scattered around that value. Then, anything outside that range or scatter can be regarded as abnormal. Obviously, larger steering angles than normal for a given cornering force indicate understeer, and smaller angles, or even 'opposite' angles such as the point at −1.20G (right turn) and about +4 degrees steering angle (left lock) indicate oversteer (in this instance, a big steering correction was necessary). Again, the ability of this particular DAS to link the track map with points of interest on graphs can be used to indicate where on the track these events occurred. Thus, the data at the bottom left of Figure 11-4, which seem to point to spates of understeer, may well happen at particular corners, or phases of corners on the track. Such a plot can therefore be used to quickly home in on balance or driver technique problems.

X-Y plots, as we have seen, are useful ways of examining the relationship between two parameters, both to analyse the pattern of the relationship, and to spot variations and anomalies that do not follow the general trend – both are useful analytical techniques. We have only looked at a few examples of the applications of X-Y plots. It is possible to envisage others, such as oil pressure versus rpm, or exhaust gas temperature versus rpm, if you were able to monitor those channels. In fact, the more channels you do log, then the more possibilities there are to look for the relationships between them. It also becomes possible to make X-Y-Z plots, in which three inter-related parameters can be viewed at once, but we'll leave such niceties aside in this book to concentrate on ways of obtaining maximum value from your DAS in the least time. If you have the occasion to study a problem involving three inter-related parameters, don't forget, an X-Y-Z plot may help!

Maths channels
The more sophisticated and expensive DAS systems come armed with a set of preloaded maths functions that allow you to perform calculations on your recorded data, and plot them as if they were recorded channels. So, for example, if you decided that the square root of the throttle position multiplied by the cosine of the steering angle, all divided by how many bananas the driver ate for lunch was going to tell you something useful, then you could enter the formula through your software, and have the resulting mess plotted out as one of your channels.

I don't propose to go too deeply into maths channels here, for two reasons; first, my maths isn't up to it; and secondly; much of what you can find out through data logging can be achieved by keeping things as simple as possible. Data logging is a tool that supplements the usual process of thinking and talking. Don't let it take you over, unless it's your job! Even in that situation, keep in touch with what's going on in the real world by talking with the drivers and engineers! But, a couple of examples of maths channels will help to illustrate their potential usefulness.

We discussed earlier how steering angle and cornering force are related. But we actually took a simplistic view of that relationship, because if you stop to think about it, the steering angles you get in fast corners, which by definition are large radius corners, will be less than those you see in low speed, tight radius corners. So, could a mathematical correction be applied, I hear you ask?

One suggestion has been to generate a corrected steering trace by multiplying the steering angle by the square of the speed, since cornering radius is proportional to the square of speed. But this apparently exaggerates steering angle at higher speeds, and an improved correction is said to be to multiply the steering angle by (speed x √speed). The resulting 'speed adjusted steering' channel should then show consistent corrected steering angles at all speeds, unless a handling imbalance is present at some particular part of the speed range. So, for example, the corrected steering channel would hopefully show up excessive steering angle being applied in fast corners if a car generates high-speed understeer, but is nicely balanced at low and medium speeds.

Another example of a maths channel that is intended to point up the occurrence of oversteer or understeer is given by a more complicated formula, again involving speed and steering, and also lateral G; a so-called OSUS (oversteer or understeer) indicator is defined thus;

$$OSUS = [\text{steering angle} - (\text{wheelbase} \times \text{lateral G} \times \text{constant})] / \text{speed}^2$$

The wheelbase and the appropriate constant will vary from vehicle to vehicle, but the purpose of the equation is again to try to determine whether the steering angle is greater or less than it should be at a given speed in a given corner. My feeling is that studying the data trace shapes and talking to the driver (if that's not you) will probably get you to the answers you want with greater reliability than the best of mathematical formulae, but then I'm not a mathematician. Any mathematicians reading this will have just called me a half-wit and closed the book in disgust. Just use the tools with which you're most comfortable and confident.

An alternative to the built-in and user-defined maths channels in the expensive packages is to export the data to a spreadsheet and do your own calculations and plots there. The preceding sentence is, of course, heavily loaded with computing jargon, which I promised I wouldn't do. However, if you know how to use spreadsheets (the computer software equivalent of a programmable calculator that can also create graphs for you), then you will need no further explanations from me on how to go about extracting the raw data from your DAS software, cutting and pasting into a spreadsheet, and then applying your own graphical and mathematical analysis to it. Apologies to the computing starters, but there just isn't the space in this book to expand on that last load of gobbledegook. But there are plenty of excellent books already about using the widely available spreadsheet packages like Microsoft's Excel and Corel's Quattro Pro, some even written in plain and understandable language.

So, even if your DAS analysis software does not include the ability to create maths channels, you could perhaps try exporting the raw data, either by using the mouse to cut and paste the data into the spreadsheet or, if there's no alternative, tapping it in manually via the keyboard. You will then be able to create your own X-Y plots, histograms and maths channels. But as I said just a short while ago, concentrate on getting maximum benefit from your DAS in the least possible time, because there is never enough time to analyse all the data we collect.

As if to simultaneously contradict and emphasise that last statement, the next chapter will take a brief tour of the many other sensors and channels you could have available, if you have the budget of a top pro team!

Chapter 12

Yet more channels?

THERE IS GENERAL agreement amongst DAS users that you can make big gains in information gathering, and make reasonably confident analytical diagnoses, with the six channels that have been covered in some detail so far in this book. But inevitably, the more questions you answer, the more new questions arise, and in your thirst for knowledge you are very likely to want to log even more channels once you're up and running.

Minor problems like having insufficient time to analyse the information you already bury yourself in are unlikely to deter you!

Even the basic loggers can handle quite a few more channels than the six 'main' ones we've discussed, offering endless opportunity to create your own data mountain. Top professional systems such as might be used in the world's premier racing categories can offer the

A linear potentiometer senses the displacement of this Formula 3 'monoshock' front suspension.

ability to log over a hundred channels, and we'll take a brief look at how this data is gathered and dealt with in the next chapter. Here, we'll just do the quick tour of what's available on the sensor front, and touch on the types of analysis that you might be able to get involved in.

Suspension movement and loads

After the six basic channels, suspension movement sensors are probably the next most often-used sensors in data logging. Damper position linear potentiometers (damper pots) are commonly seen on cars from the well-heeled club racing categories and upwards. Looking like small telescopic dampers nestling beside the actual spring/damper units, these devices faithfully mimic the suspension movements and send signals to the DAS unit.

In Chapter 1 the question of DAS sampling rate was touched upon, particularly in reference to logging suspension movement. It will be fairly obvious if you have ever watched a spring/damper unit moving on any type of competition car that it can do so very rapidly at times when going over bumps, kerbs, rocks and so on. It might seem a contradiction, but some of the highest damper piston velocities occur on vehicles with soft springing, but long travel – rally raid vehicles are possibly the most extreme example in four wheeled competition, but rally cars also generate fast suspension movement. Speeds of over 300mm/second are possible in this type of situation, and so a fast sampling rate is essential to capture the data with any degree of precision. But in other categories where aerodynamic downforce is low, and hence spring rates are lower and suspension travel is greater, high damper piston speeds are also encountered. So even if you flog around your local circuit rather than from Paris to Dakar, and you want to record the detail of suspension movement then it is generally reckoned that you need to log at no slower than 100Hz, preferably at 250Hz, and at 500Hz, if your system will allow it.

Logging suspension movement can tell you how much the suspension at each corner of the car moves and how fast it moves in response to short-lived, 'high speed' (that is, suspension movement speed) events like bumps and kerbs. It also tells you how much it moves and how fast during lower frequency 'slow speed' events like roll and pitch, and how much the suspension moves in response to aerodynamic loads at varying speeds. All of these capabilities can permit the knowledgeable race engineer to fine-tune the springs and dampers, spot and diagnose anomalies in handling, and check on aerodynamic balance.

It is common practice to fit damper pots to all four corners of a car – so that's another three or four channels of data just to plot the suspension displacement curves measured directly from the damper pots. If you then use the DAS software (if it is thus enabled) to differentiate the displacement data to generate a damper velocity plot (sorry about the mathematical mumbo-jumbo, but a reasonably sophisticated DAS does this tricky stuff for you anyway), you effectively create four more (maths) channels too. But you also gain an idea of the actual velocities your dampers are dealing with, which can help you zero in on the best way to tune them. Dampers, remember, are velocity-sensitive devices, and many now permit the separate tuning of the high and low piston speed responses. So discovering the actual piston speeds encountered on a given car, and even on a given track or corner, can be of great help if you know what's needed to tune the dampers to match.

Suspension loads can also be logged. Through the use of strain gauges bonded directly to suspension members like pushrods, or load cells (whose output is based on internal strain gauges) mounted co-axially with the pushrods, the actual loads encountered by the suspension can be measured dynamically (and statically come to that). This enables dynamic weight transfer during roll and pitch changes to be analysed as well as the peak loads developed on bumps and kerbs, not forgetting the additional loads imposed by aerodynamic downforce (or lift if your category doesn't allow downforce-inducing devices).

Data on suspension loads can be co-plotted with damper displacement and velocity data to present a pretty full picture of what happens to the car's suspension and the basic aerodynamic platform (with respect to changes of ride height and rake). Figure 12-1 is taken from a session with a Formula Renault Sport single seater at Croft circuit in 1997 (from the same session as has been used for earlier examples we've looked at). In such a small representation of the full computer screen it's not easy to see all the detail here, but the 12 traces shown represent four sets of data, one for each corner of the car. Each data set consists of measured damper displacement, a calculated damper velocity trace, and measured (by strain gauges on the pushrods) suspension load. The traces start at the top of the screen plot in Figure 12-1 with the front left velocity (FLVEL), load (FLSG) and displacement (FLDISP), then the same data for the front right. In the lower half of the screen, in slightly different order, is the data for the rear of the car, going from rear left displacement (RLDISP), to rear left velocity (RLVEL), to rear left load (RLSG), and then the same data for the rear right. Notice that the x-axis is graduated in time (seconds) rather than the more usual distance. That's because of the need to calculate velocity from displacement data, which obviously involves reference to time. It is possible to view this data against distance, but it is less clear and less meaningful, especially when you zoom in to finer detail.

Immediately you can see that there are high frequency, short duration events here, going by the spikey nature of the traces, but also longer duration events where wider peaks and troughs, and longer slopes occur. You can see that

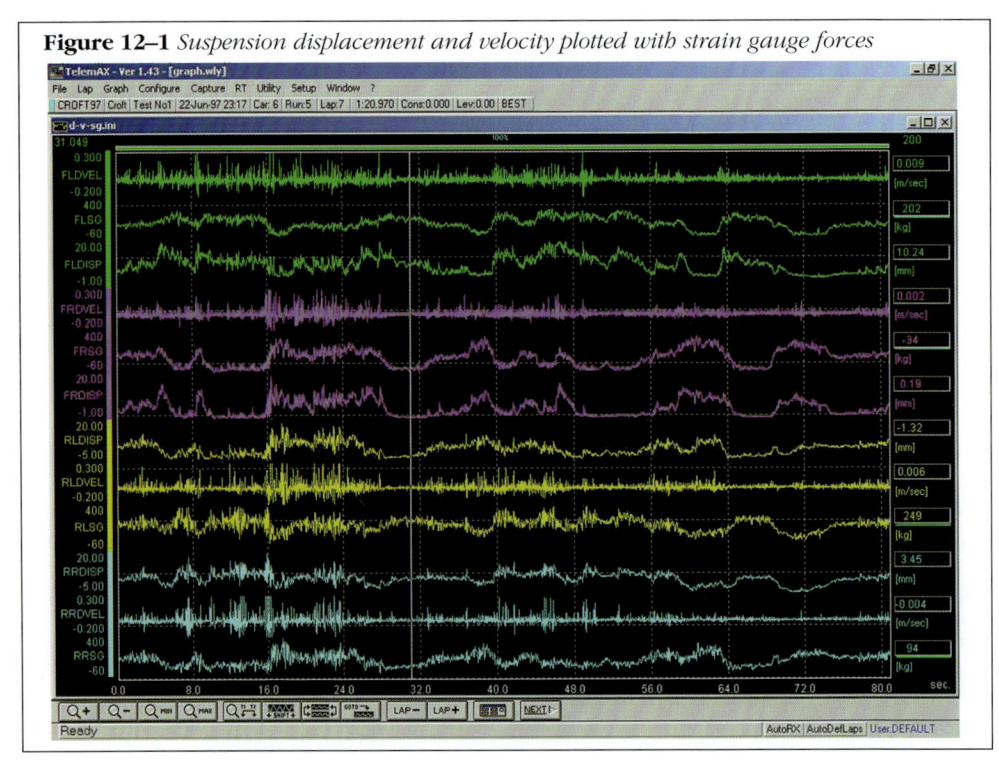

Figure 12–1 *Suspension displacement and velocity plotted with strain gauge forces*

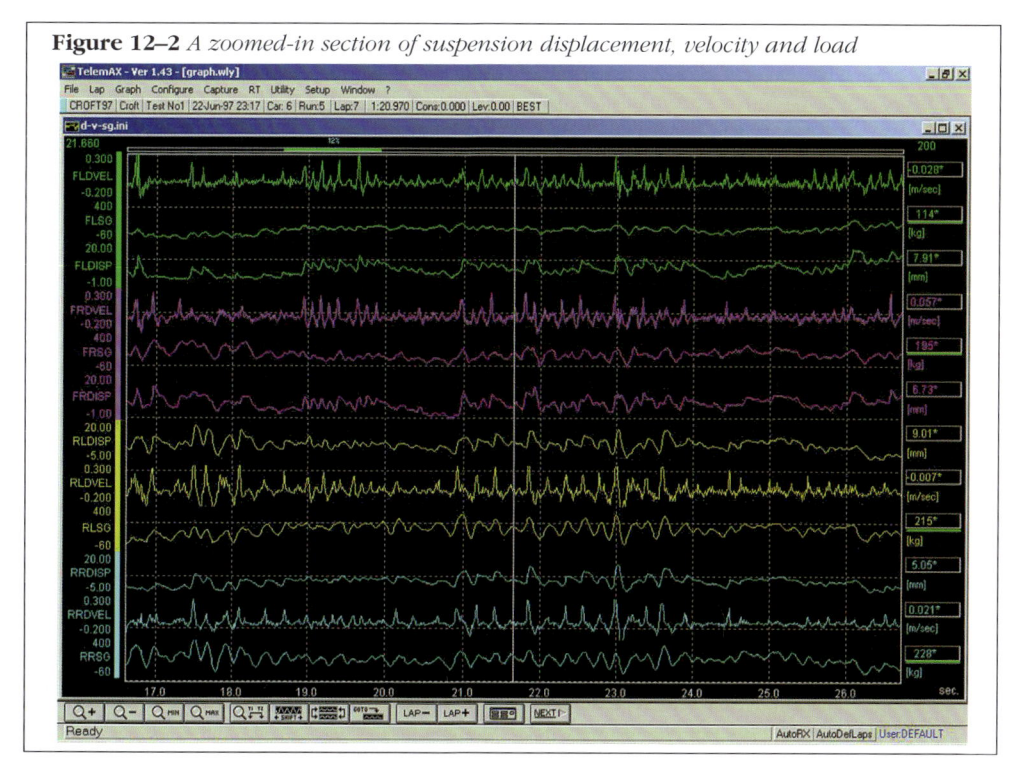

Figure 12–2 *A zoomed-in section of suspension displacement, velocity and load*

where a rapid displacement occurs, a velocity spike and matching strain gauge load peak also occur. You can see events where both front wheels show similar displacements (two-wheel bumps), such as around the six to eight seconds region. Also, there are events where the front wheel displacements are completely opposite (due to roll), such as occurred around the 60-second mark. Figure 12-2 shows a 10-second chunk of this same lap zoomed in to expand the short-duration events. It is now easier to see how the load peaks correspond to the velocity peaks, for example.

Once this kind of data is available it might become clear, for example, that you are dealing with a bumpy surface, which makes the suspension and dampers work more in the high-velocity regions, and you therefore might have to focus your attentions on tuning the high speed responses of your dampers to optimise handling and grip. On the other hand, it may become apparent that the surface is smooth and relatively bump-free and that you are therefore perhaps going to be better employed working on the low-speed valving, if you use dampers to tune the car in this way – top single-seater race engineers are veering away from using dampers to tune the car for changes in roll and pitch, preferring adjustments to springs, anti-roll bars, bump rubbers and anti-dive and anti-squat geometry, especially on cars with significant aerodynamic downforce. But that's another story.

The main point to stress here is that by logging this kind of data you don't have to guess at the displacements, velocities and loads your suspension is dealing with – it's all there in front of you to study and analyse.

Ride height sensors

Clearly the suspension movement sensors can be used to get an idea of dynamic changes of ride height, especially if you take the trouble to either calibrate your damper pots as ride height sensors, or you manually measure the ride heights corresponding to the damper positions on full compression and full droop in the workshop. However, this does not take into account the effect of tyre deflection, a problem that becomes more relevant when substantial downforce is produced and the tyres' spring rates are a significant part of the springing of the car. But it is possible to measure ride height dynamically and directly using laser ride height sensors.

In very basic terms a laser ride height sensor bounces a low power laser beam on to the track surface, and the point at which it is reflected back is used to determine the distance to the ground. Ride height variations of as little as 0.2mm are apparently discernible, and Champ Car and F1 teams, who rely heavily on optimising the aerodynamics of their cars' underbodies, have been able to refine their set-ups using these sensors.

It is possible to use a pair of ride height sensors positioned across the car to determine and refine not just chassis roll, but also the sideways tilt on a car running on ovals – the cars are set up with a static tilt that is supposed to then settle flat for the best underbody airflow and hence the best aerodynamic grip in the banked corners.

Ride height sensors can be fitted to wheel uprights to enable tyre deflection to be measured directly. On anything other than a perfectly smooth surface the data would be dominated by the ups and downs of the wheels as they rode over bumps, but on smooth surfaces, the tyre compression at speed due to aerodynamic loadings could be analysed this way.

Aerodynamic sensors

Reference was made in the foregoing sections to the direct and indirect measurement of aerodynamic forces using displacement, load or ride height sensors. These measurements can undoubtedly provide valuable information about the levels of downforce (or lift) acting on the suspension, or the displacement of the chassis, as the result of these aerodynamic loads.

Thus, it becomes possible to make changes to the aerodynamic set-up of a competition car and actually measure the changes to the aerodynamically induced forces, to the front-to-rear distribution of those forces, and to the ride-height and rake of the chassis. This data can then be correlated with other logged performance data such as speeds in corners and on straights, and can also be cross-referenced to data related to the car's handling to see if the chassis balance changes in response to measured shifts in aerodynamic balance.

As an example, increasing the angle of attack of the front wings will lead to more downforce at speed on the front wheels (measured by strain gauges or load cells), probably also a reduction in front ride height (measured by displacement sensors and/or a ride height sensor), and if high speed understeer was previously present then it should be reduced or eradicated. Thus, reductions in steering angle in the high speed corners might also be expected to follow such a change.

This is good, basic track-related information which is invaluable in establishing a database of aerodynamic settings and their effects on handling balance, cornering ability at given speeds, and straight line performance (related to aerodynamic drag). But for research and development on a more academic basis, aerodynamic pressures can also be measured and logged.

Downforce is created on competition cars by the manipulation of the airflow to produce pressure differentials between the upper and lower surfaces. Thus, reduced pressure beneath a car holds it to the ground more firmly and increases grip. Similarly, chassis-mounted aerofoils, or wings if you prefer, generate pressure differentials between their upper and lower surfaces and push the car more firmly on to the track, again to increase grip.

But making these airflow manipulation devices work efficiently is paramount to optimising performance. Downforce needs to be generated that produces the smallest possible increases in drag (you rarely get the one without the other). This means managing the airflow with carefully crafted shapes that not only create the desired pressure drops, but which also return the air pressure back to ambient levels as efficiently as possible.

The ability to measure these relatively small pressure differentials, and to be able to map them over the surfaces of downforce-inducing components (or indeed entire car body surfaces), is therefore a useful part of competition car aerodynamic research and development. Such measurements are, of course, made on a more or less continuous basis in the wind tunnels of the top teams nowadays, and the top data acquisition companies are involved in the instrumentation of many race teams' wind tunnels.

But it is also possible, and necessary, to log pressure data on full-size competition cars out on the track. It is necessary for the top teams to do this because the data derived from testing scale models, even ones that are as big as half full-size, is not exactly the same as on a full-size vehicle, and practicalities currently preclude testing in wind tunnels at full size for most. It is also necessary if aerodynamic data is deemed essential but the team does not have ready access to a wind tunnel, or a model to use in it.

Thus, the data acquisition companies have come up with pressure sensors that can either be used singly or in multiples. Pi Research for example has single aero sensors and also a hex aero sensor which contains six individual low-pressure sensors in one unit. The sensors can either have a measurement range of +/− 0.5psi or +/− 2psi, and are connected to sensor ports in the surface of the car or wing under investigation via plastic tubing. It then becomes possible to produce plots of surface pressure beneath sidepods and diffusers, or on the surfaces of wings, and from that deduce whether the pressure distributions need changing with alterations to shape and design.

Other aero sensors include the pitot tube, a device which measures air velocity as distinct from wheel speed (in other words it takes any wind speed into account), and other air speed sensors, based on mini-propellers, are also available.

It could perhaps be argued that a car equipped with load cells and aero sensors is a reasonably cost-effective alternative to a wind tunnel programme, at least in terms of capital investment if not in running expenses.

Tyre pressure and temperature sensors

Pi Research markets its HP Tire Monitoring System (HPTMS) as a means of providing early warning of potential tyre failure by measuring inflation pressure. The system allows data to be logged, but of course it is only of any use as an early-warning system if the data can be transmitted via telemetry directly to the pits for constant monitoring. Thus, I'm getting a bit ahead of myself because telemetry isn't due to be covered until the next chapter. However, logging tyre inflation pressure could be perceived as useful in more than the safety context, so let's take a peek anyway.

The TMS consists of a car-mounted radio receiver and four wheel-mounted sensor/transmitter assemblies. A wheel sensor/transmitter fitted to a Lola Champ Car is shown in the photo.

A tyre pressure sensor on a Champ Car.

Specially modified or manufactured wheels are required for this system, and the modifications must be done by the wheel manufacturers and approved by the relevant motorsports regulatory body for any given category.

The sensor/transmitters bolt through the wheel rims, and each is powered by a pair of lithium batteries. Although the body of the device is outside the tyre, the 110mm long antenna is actually inside the tyre. Each sensor/transmitter weighs about 85gm, and to offset this weight they are mounted opposite the valve and a small counterweight is also fixed to each wheel. Data transmission from the sensor to the on-car receiver only commences when wheel motion is detected, then inflation pressure data is transmitted every 20 seconds (although readings are logged more rapidly). If the pressure changes by + or − 0.5psi the data is then transmitted every 2.5 seconds or until the pressure stabilises. Each receiver is programmed to accept data only from the four sensors within a given system, and similarly each transmitter has a unique programmed identity so that teams cannot receive data from the wrong car. The TMS requires one of Pi's more sophisticated data logging systems to run it.

At the time this book went to press, Pi Research was working on an update of the TMS that will log and transmit data more accurately and more frequently, this system being intended to cater for chassis and handling analysis as well as performing the safety role of its predecessor.

Tyre temperature is obviously another critical performance parameter. This too can be logged with different types of non-contact infrared temperature sensors, and by utilising three (or more) such sensors across the width of a tyre, a temperature profile can be logged for later analysis. Such data is of great use when tuning and optimising the suspension settings.

Counting channels again, the TMS alone constitutes four channels, but if supplemented with three tyre temperature sensors per tyre that would add another 12 channels.

Brake sensors

Under this very general heading come brake line pressure, disc temperature and pad wear sensors.

Brake line pressure was mentioned during our discussions of longitudinal G, and can be thought of as a more direct measure of the driver's input to the brake system than long G, which is the final result of the use of that system. But equally, knowing that the driver is generating 2,000psi, or whatever, in the brake lines doesn't really tell you anything of great value. Knowing that he generates - 2G under braking when you believe he should be attaining −3G is, however, of value. But there may be occasions, for example when establishing a front-to-rear brake balance, that a logging brake line pressures will be of value.

Temperatures of brake discs depend fundamentally on the weight and speed reduction of the car, but also on disc and pad materials, the demands placed on them by a given circuit, any cooling provided, and driver technique. Logging the disc temperatures will obviously provide the best understanding of the temperature range that the discs go through, and will thus enable the most suitable choice of materials to be made.

Brake cooling is also a function that can have a very significant effect on a competition car's aerodynamic performance. Thus, the ability to log the effects on disc temperatures of alterations to brake cooling can clearly be very useful.

Pad wear is probably of most significance in endurance racing, but is also of some importance in relatively short Grand Prix races, where pretty much every component is designed and manufactured to last the race distance, but no more! Mentioned in Chapter 1 was the linear variable differential transformer, or LVDT, which is used for measuring small position changes or displacements where conventional linear potentiometers are too insensitive and bulky. LVDTs have been used with success at Le Mans, where pad wear could be logged and transmitted via telemetry to the pits so that replacement intervals could be accurately estimated and pit stop strategies optimised.

Engine systems

There are many more sensors that have found use on competition cars. Of these, we have barely mentioned all the engine system monitoring sensors that are utilised on gas and fluid temperatures, pressures and flow rates. These include water, oil and fuel system temperatures and pressures, and not only does logged data from these systems provide engine builders with a fairly complete history of the work the systems have been doing, but they also provide the requisite information for system refinements.

For example, if by studying water temperature data throughout a season, and comparing it with ambient temperatures noted or recorded at each event, it becomes clear that your cooling system never allows the water temperature to rise above 65°C, you might reasonably conclude that you could reduce the size of the radiator matrix. This in turn might allow you to reduce your car's aerodynamic drag, which could then give you better straight line speed next season.

Oil system gremlins that you wouldn't spot whilst driving can also be picked up too. For example, oil surge that allows air, albeit briefly, to be sucked into the oil system from an under-baffled sump might well cause engine-damaging drops in oil pressure. Logging oil pressure at any reasonable sampling rate will pick up any short-lived dips and quite possibly save a lot of expense.

Fuel pressure can often drop in line with battery voltage and cause hard-to-find misfires and erratic engine running. Logging battery voltage alone could well point you in the direction of the root cause of a misfire even if you are uncer-

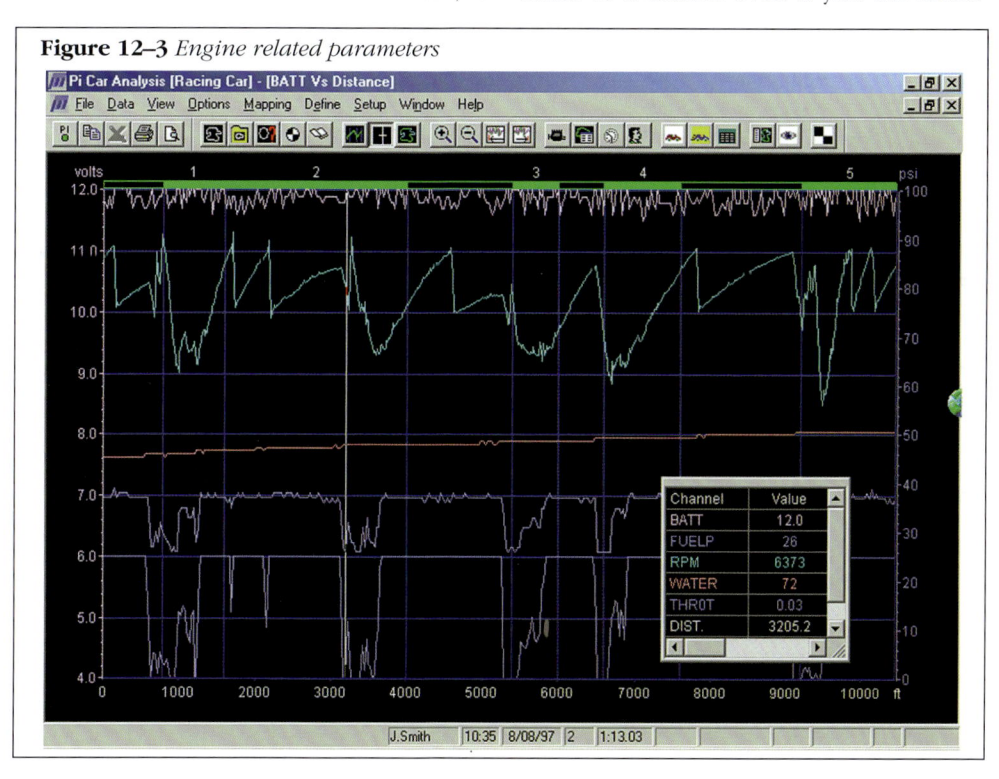

Figure 12–3 *Engine related parameters*

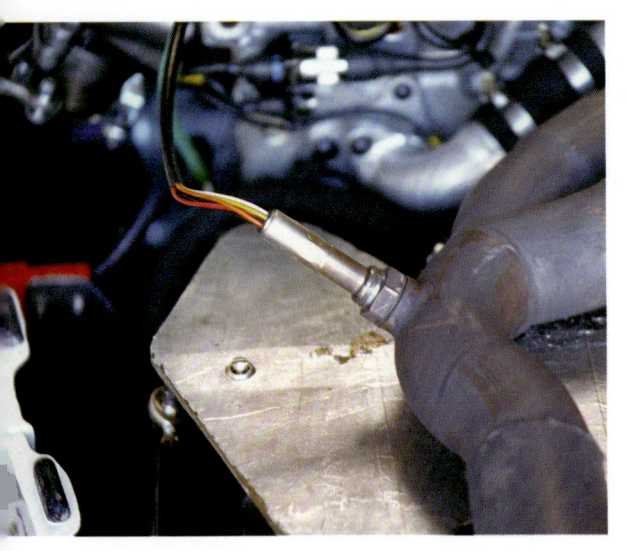

An exhaust gas temperature (EGT) sensor.

tain as to whether poor fuel pressure or a weak spark was the apparent cause. Logging fuel pressure too might well help to eliminate one factor in the search for a misfire. Figure 12-3 is an example of a logged set of engine-related parameters.

Exhaust gas temperature is another engine parameter that can usefully be logged, along with temperatures of the engine fluids at various locations.

A shaft torque sensor kit. (Courtesy Pi Research)

Torque sensors

Another specialised application of strain gauges is the torque sensor. Pi Research again markets these, and in essence they comprise strain gauges fitted to a drive shaft or other shaft in the driveline in a way that detects the minute amount of twist that occurs when the shafts transmit torque. The clever part about the packaging of the Pi Torque Sensor system is that it allows data to be logged while the shaft is in use, and therefore rotating. This is done by using a lightweight transmitter which clamps to the shaft in question, and a car-mounted receiver dedicated to that particular transmitter. Kits are sold with left- and right-hand transmitter/ receiver sets.

If there was ever any doubt as to how it is possible to end up logging over a hundred channels, it must now be obvious how the total channel count could reach large numbers – providing the budget was available that is.

Pi VIDS

At the risk of seeming to favour one organisation's products (it has to be said that Pi Research IS a very active manufacturer in the field), one particular new

Figure 12–4 *'PiVIDS', or 'Video Indexed Data System'*

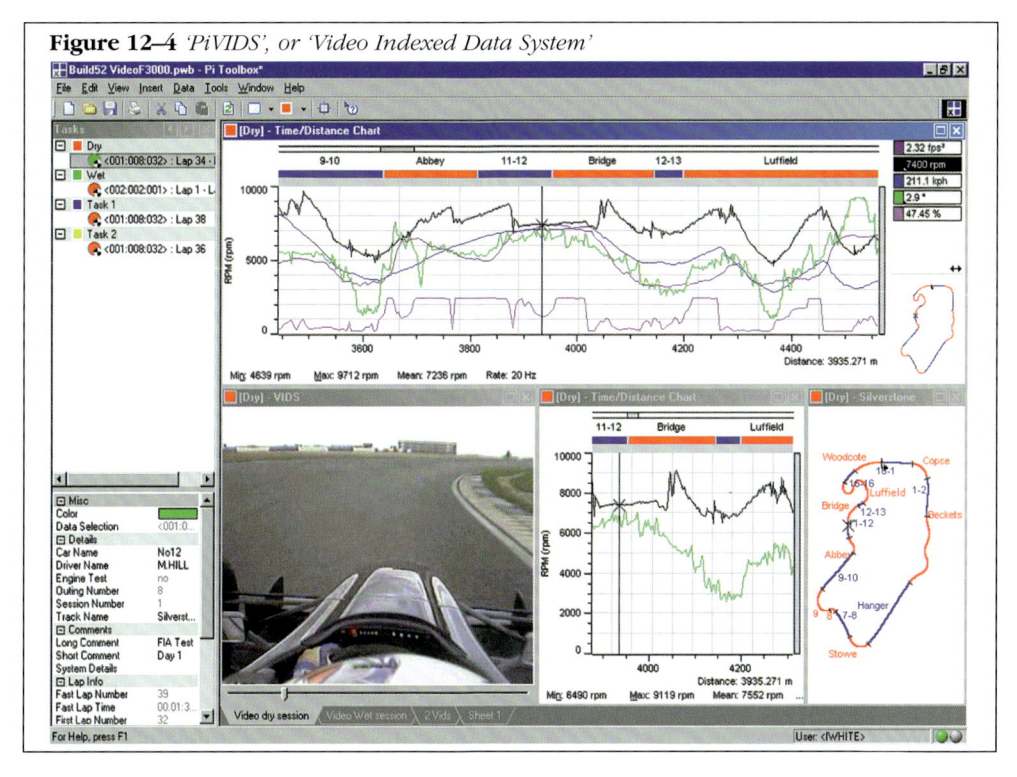

product (as this is written in late 2001) worthy of special note is Pi VIDS, an acronym for its video indexed data system. With the aid of the camera included in the kit, video images are recorded on removable memory cards. The memory card can then be removed from the car and played back using Pi's V6 Analysis or Pi Toolbox software, and the video image is synchronised and displayed with the logged data (see Figure 12-4).

This kit has been finding its way into F1, F3000, and F3 as well as the main US racing categories. Its greatest value is perceived at present in being able to see what drivers are doing in the cockpit, the lines being taken around the circuit and so on, and it is helping to bridge the communications gap between race engineers and drivers. But the camera provided is small, light and rugged, and is capable of being attached almost anywhere on the car. Already some fairly fundamental discoveries have been made by pointing the camera at seemingly strong, stiff suspension members and aerodynamic devices . . .

You may ask why you need a data logger to get the benefit from using video in this way. Probably you don't, but the ability to synchronise the video footage with the data makes it that much more readily useable and valuable, and it does the job that the race engineers would want it to do – allows them to see at a glance where on the circuit the video-recorded events took place. Sound tracks add a nice touch too, and can provide an aural clue as to what the driver was doing at any one time.

Typically, in the States one of the ways that PiVIDS is finding application is in selling videos to pupils at track experience days as a memento of their fun day out. What a great idea – and a lot easier to impress your friends with than just a data logging file or printout!

Chapter 13

Data logging in use

THE SET OF examples of data logging in use picked and laid before you here is by no means a comprehensive selection, but hopefully it reflects the use of different levels of sophistication in a few different competition disciplines. Some of the examples relate to the personal experience of the people involved whilst one is necessarily general in nature, as you'll see when you get to the section on Formula 1. We'll kick off with my own introduction to the topic at the bottom end of the sophistication scale.

Rpm logging in hillclimbing

Rpm logging was covered in some detail in Chapter 4, but a few personal notes might help to emphasise the benefits I discovered on my introduction to this most basic of data logging systems. I was sharing a 2-litre Hart 420R-powered Pilbeam single-seater hillclimb car with my brother Andy, which we ran jointly from 1987 through to the mid-'90s. The car was pretty well developed and competitive by the time we installed the requisite recording upgrade into our Stack

The author's early data logging experience involved a Neolithic laptop! (Tracey Inglis)

tachometer and purchased the analysis software, the price of which Stack had recently reduced from its earlier ridiculous level to something much more sensible.

The tacho logged rpm at 10Hz and had a 400-second memory, about as basic as it gets, but to pack the runs into the small memory space we simply fitted a separate power switch which enabled us to power up the tacho just before a run and power down just after the finish line. With (typically) four runs each of 30 to 50 seconds duration, we could usually pack a full meeting's worth of data into the memory this way. Unless and until it was downloaded the system would overwrite earlier data, like a continuous recording tape, so at worst you might lose the data from the first practice runs if you exceeded the memory capacity with longer runs.

To obtain value from the rpm recordings on the day we would usually download after our (generally two) practice runs. The software allowed two runs to be viewed on screen at any one time, enabling comparisons to be made. The fact that there were two of us sharing the car meant we could learn from each other's data, and there is absolutely no doubt in my mind that we were both able to go quicker as a result of studying the data.

There was plenty of useful information to be gleaned, as detailed in Chapter 4, which in essence boiled down to taking away the guesswork on the gearing, enabling us to get that aspect of set-up absolutely spot on, and being able to gain time by studying each other's mistakes and successes. It was indeed also possible to spot evidence of understeer and oversteer from the shape of the rev trace in particular corners, especially by comparison with the expected patterns. Often these apparent handling quirks were down to driver technique, something that is clear when comparing the data from two different drivers in the same car, but at least the evidence was there that could help generate solutions.

It might be instructive to look at the

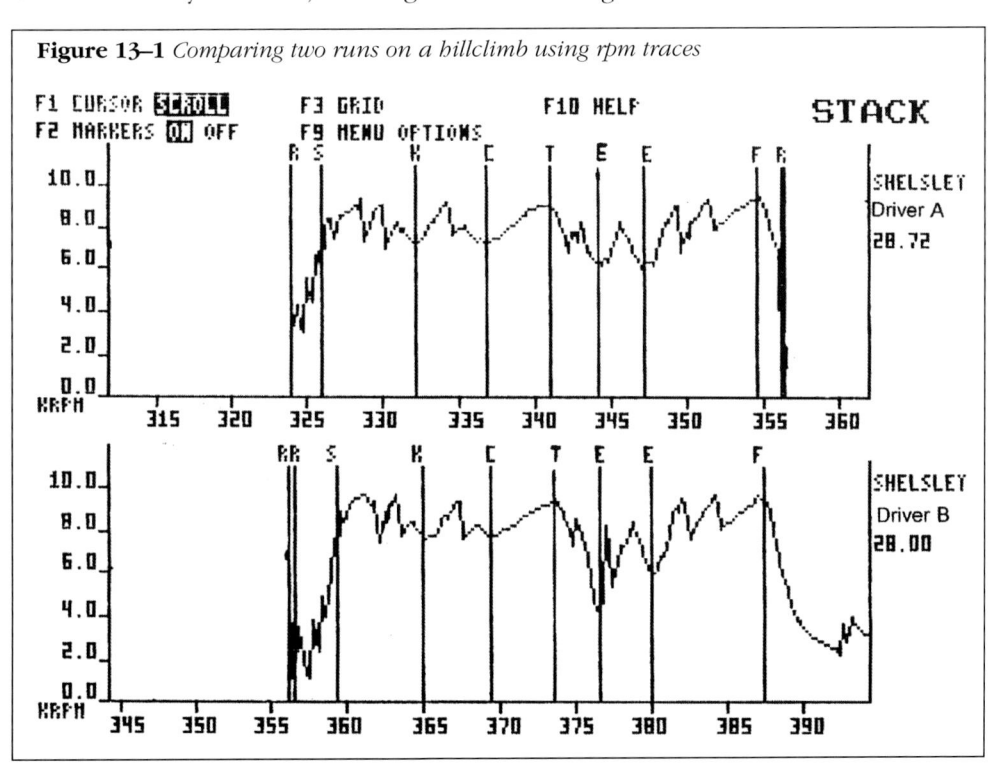

Figure 13–1 *Comparing two runs on a hillclimb using rpm traces*

data from one of the events we did in the mid-'90s, and to see how we broke it down. Figure 13-1 shows a screen plot of two runs, one set above the other, at the famous Shelsley Walsh speed hillclimb in Worcestershire, England, the course map of which is shown in Figure 13-2 (annotated with numbers showing our approximate gear change points). This 1,000yd (915m) hill is one of the faster mainland UK courses, with an average gradient of 1 in 10.42 (9.6 per cent), and a steepest section, between the Esses, of 1 in 6.26 (16 per cent). The course record equates to an average speed, from a standing start, of approximately 80mph (128km/h), with top speeds by the fastest cars exceeding 140mph (225km/h).

The vertical markers on the chart were either put in by the tacho software (R, meaning reset) or manually by us to indicate the start (S), Kennel Bend (K), Crossing (C), the Ess Approach speed trap (T), the Esses (two Es) and the finish (F). The table summarises the key differences between the runs.

Remarks	Driver A	Driver B
Start	Drops clutch at 6,800rpm Searching for traction	Drops clutch at 7,800rpm Smooth getaway
Shift to 2nd	@ 9,000rpm	@ 9,600rpm
Shift to 3rd	@ 9,000rpm	@ 9,600rpm
Lift for Kennel	1.26sec before apex	1.16sec before apex
Speed through Kennel	73mph	77mph
Time to Kennel	6.26sec	5.91sec (−0.35sec)
Shift to 4th	@ 9,000rpm	@ 9,200rpm
Lift for Crossing	1.26sec before apex	0.94sec before apex
Speed through Crossing	84mph	89mph
Time to Crossing	10.79sec	10.00sec (−0.79sec)
Speed through trap	105mph	108mph
Time on brakes	2.96sec (−0.88g)	2.70sec (−1.06g)
Speed in Bottom 'S'	48mph	46mph
Time to Bottom 'S'	18.40sec	17.20sec (−1.20sec)
Speed in Top 'S'	46mph	46mph
Time to Top 'S'	21.23sec	20.69sec (−0.54 sec)
Shift to 3rd	@ 9,000rpm	@ 9,500rpm
Shift to 4th	@ 9,200rpm	@ 9,600rpm
Speeds over Finish	109mph	110mph
Time to Finish	28.72sec	28.00sec

It is very evident that driver A, having studied the rpm trace of driver B, would be armed with a good deal of information to help shave some time off the next drive. Start technique, using the correct shift rpm, and maintaining speed through the faster corners were all areas that driver A could have improved. Interestingly, on the evidence of these two runs, driver B could also improve, having lost a significant chunk of time in the Esses. The best 'composite segment' time from these two runs was 27.34 seconds, indicating that both drivers were some way off what was theoretically possible on the day. Sad to relate that driver A was the author, and it may be no coincidence that I have more second and third in class trophies from this venue than firsts! But we did win the 2-litre racing car class at every Shelsley meeting between us during the season from whence this data came, and I'm convinced that our use of even this most basic data logging system contributed to that record.

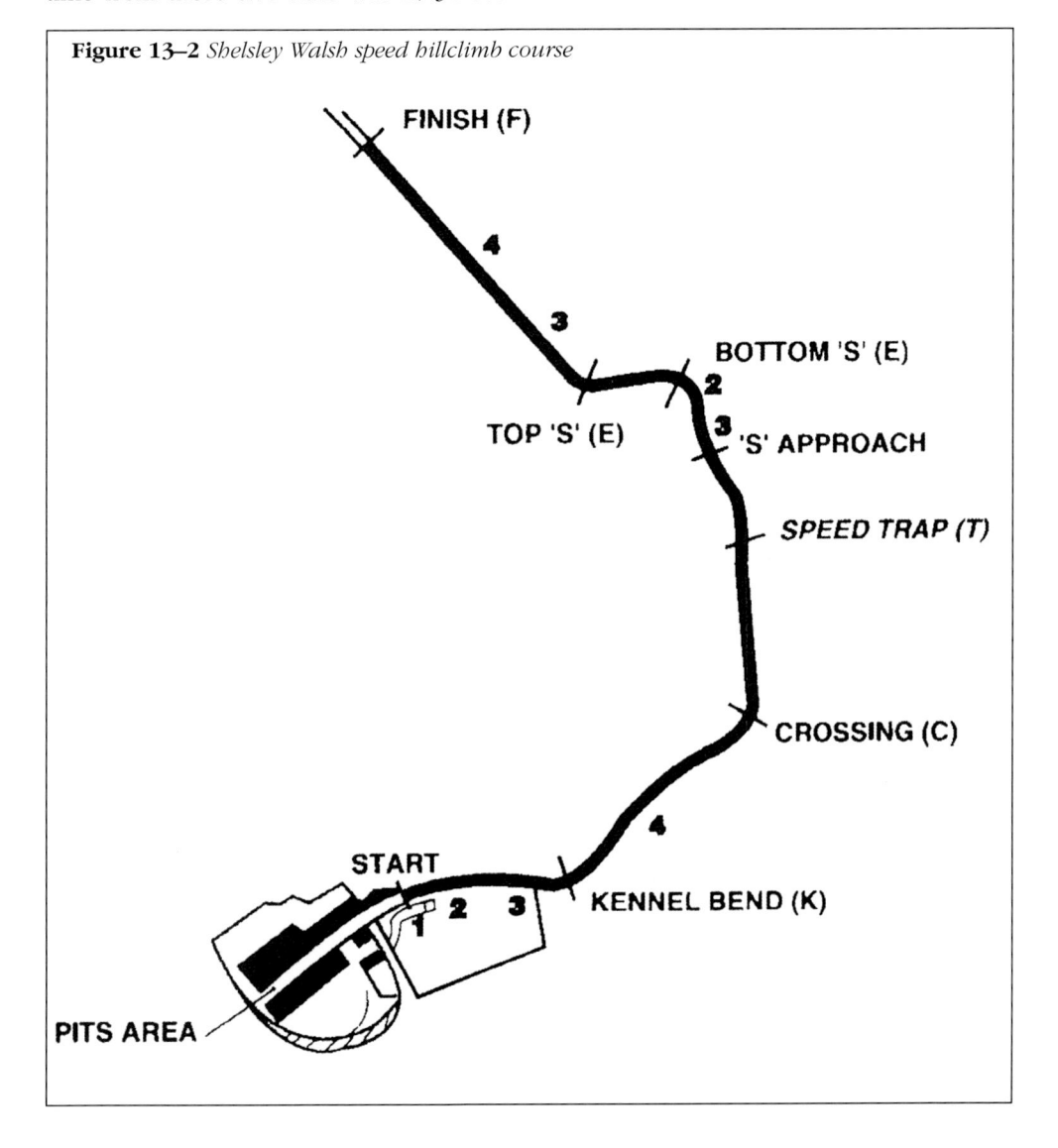

Figure 13–2 *Shelsley Walsh speed hillclimb course*

Driver coaching

Constant reference has been made throughout this book of the benefits of making data comparisons, between runs, laps and especially between different drivers. In the example above, it was very evident that driver A could benefit from the data created by the faster (on that run . . .) driver B, providing an open-minded approach was adopted. Which leads to the next example of making use of data logging – driver coaching.

Coaching is relatively new in motorsport. For some reason, probably to do with fragile egos or just plain old (and sometimes misguided) arrogant self-belief, racing drivers have been very slow off the mark to take up the benefits of coaching. Even those that have used coaching seem set on keeping it quiet, in contrast to just about every other sport where world-class competitors regularly offer outspoken thanks to their coaches. Surely to admit you need a coach is to acknowledge that you can always improve, not that you're no good in the first place.

But whatever the reason for the slow start, driver coaching is now on the increase, and the wide availability of data logging has certainly contributed to this. In Chapter 5 I referred to an interview I did with Nigel Greensall, former BOSS series champion driving an ex-Grand Prix Tyrrell. As well as competing successfully on the national racing scene in the UK and Europe, Nigel also works as a professional driver coach. You can hire his services to come and drive your car with you at test sessions, knowing that he is a naturally talented and quick driver who will almost certainly go quicker than you in your own car! Nigel's perspectives on driver coaching, the benefits that data logging bring to it, and on the approach you need to get the best out of your data logging system are worth reading more fully. Forgive the very brief duplication of a couple of excerpts already stated in Chapter 5.

Professional driver coach Nigel Greensall being listened to avidly by student Tim Christmas aboard his Radical sports racer.

Q. If you had the budget for a basic data logger, what channels would you pick, and why?

A. 'I'd always go for speed and rpm, they're the first two channels I'd pick, purely because as a driver training aid, which is where I find data loggers are very, very useful, when you do graph overlays they're the two that are easiest to understand. It's easy for people to get a picture in their own mind, where they are on the circuit, and what they're doing.

'Quite often in terms of driver training it's the basics like gear changing where a lot of people are losing a lot of lap time, and looking at the rpm traces you can tell exactly how they're doing it.

'Throttle position is probably the third best. Lots of people talk about going flat through corners and, you know, the throttle . . . isn't completely flat, let's say, through the corner. But also some people are too aggressive on the throttle, maybe getting on the throttle too soon in the corner. Maybe they've lost too much speed on the entry to the corner, and then they use too much power too soon. That can create understeer on a car and that can cause problems, so being able to see the throttle movement is very useful.

'And then after that you'd probably look at either brakes (via longitudinal G) and steering. I find steering a useful one. Sometimes people get very busy with the steering wheel, and they need to slow the steering down, almost steer in slow motion. Once you can overlay one driver's data with another, then you can see the differences.'

Q. The ability to overlay comparisons is probably one of the best things from the driver training standpoint?

A. 'I think it is for any team, where they've got two cars with two drivers, or endurance racing where you've got one car with two drivers, or where the drivers are trying to improve and they're hiring in a driver coach. They can get some useful data where somebody is producing quicker lap times than themselves, and they can also see the reasons why. They're absolutely wonderful driver training aids.'

Q. What about if you're looking for the symptoms of understeer and oversteer, does that have to tie up with driver feedback?

A. 'That's difficult really because handling characteristics of cars are always induced by the driver. And so whichever way you tune a car, if you put a different driver in it he can produce different handling characteristics. Going from extremes you can put driver A in the car and he'll produce understeer and put driver B in and he'll produce oversteer, and that's not related necessarily to how the chassis is set up. But where the data logging can be useful is in actually understanding why driver A understeers through a corner and why driver B oversteers through the corner. It might be because driver A is a pillock and driver B isn't! One of these days hopefully the data logging will come up and say 'Pillock!' on the dashboard when you complete a lap!

'But from experience you can generally find out if the car is not handling properly. If, say, you've got a throttle position sensor, and the driver is reporting understeer, maybe he's come on to the throttle too soon, the understeer's come in, and he's had to lift off the throttle to recover the front of the car. By discussing it with the driver you can then interpret it correctly from that, and yes you can work out what's understeer and what's oversteer.'

Q. Do you think lateral G or steering logging help?

A. 'They build a stronger picture. In a perfect world if you've got a driver that's very honest and can give you solid feedback you can use very basic logging just to confirm and see whether there's improvement. But sometimes if a driver hasn't been able to give very solid feedback then obviously when you've got lateral G that can then become extremely useful because you can then see what sort of cornering forces he's pulling and at which points in the corners. And obviously with longitudinal G in terms of braking that can be extremely useful because quite often in a racing car people will over-brake and lose too much speed on the entry to a corner rather than relying on the mechanical or aerodynamic grip the car may have.'

Q. Do you think that the sophisticated systems can produce an overload of information?

A. 'Yes. That's why I think speed and rpm can be enough. I've always thought with racing if you keep it simple you'll do well. If you get over-complicated you'll just get lost. The simpler the better I always think. It's one of the reasons these Radicals (the cars Nigel was driving that day) are so good, they're very simple. I've driven lots of the cars now doing tuition and Radical produce a very basic set-up sheet. Other than going softer on the dampers just because I prefer a softer car I keep everything else the same, I don't change it for any circuit, and just get in and drive it. You're always on the pace and so many people go down different routes on set-up instead of looking at the data and trying to think, for example, "can I try and brake five metres later because that'll gain me half a second" or whatever.'

Q. What do you think of one driver coach's view that there's only one way to drive a car quickly because chassis dynamics dictate that there's only one way to get a car quickly through a corner?

A. 'I totally disagree with that because I don't believe there are two drivers in the world who drive the same. Say with two drivers, one is slightly taller than the other, sitting slightly higher in the car or slightly lower, what effect on aerodynamics does that have? What effect on his view of the circuit does it have? There are no two people that use the throttle the same way, or the brakes the same way. I feel that there are many different ways of driving a circuit to produce wonderful lap times and to get what is the best out of that car for that driver.

'And in all the driving tuition I've done, having driven their car I've explained to people how I've approached driving it and how I've actually driven it, and been completely open with them. It's good for them to try to copy that to a certain extent. But if ultimately their style is going to be different then you have to give them the freedom to explore what's best for their style. You'd have to say when you look at Prost and Senna for example, that they had totally different driving styles but on a given day they'd produce almost identical lap times, depending on whether the car suited one or the other more to a certain extent. But who's to say that Senna was better than Prost? It's the same with Schumacher, he's got a different driving style to a Prost.

'So I feel that each driver should be given the freedom to explore his ability to try to get the best out of his car in a way that suits his style. I know for a fact that my style's the best, but you know, it's nice that everyone else has a chance! No, I totally disagree with that viewpoint. From my point of view I always look at what race results someone's getting at the time, and really the best driver coach in the world at the moment is Michael Schumacher because he's the fastest man out there. And I always reckon that when I'm coaching somebody, unless I can get in their car and go quicker than them, then they shouldn't be paying me. Simple as that. Because if I can't go quicker than they can in their car I can't teach them anything.

'The best drivers from a coach's viewpoint are the most open-minded ones. I've worked with some people in the past who are like a closed book, they'll hire you to do some tuition but they don't really want to know, and they're not open-minded to see how they can change their style. And they'll always make excuses as to why they're slower. They'll say: 'Oh I caught traffic on that lap' and everything like that which is complete bollocks because you'll look at the data and it doesn't wash! I don't go for any of that but a lot of guys I've worked with have been really very good, very open-minded and they're willing to put their heart and soul into learning how to improve.

'The increase in data logging has made driver coaching more relevant because without data logging the driver coach can only do so much. While it's nice to sit alongside somebody in a road car and give them tuition and coaching, that's only relevant to a very small degree with what you're actually going to do in a racing car. It is relevant, and it's good to do but you also need to be able to do it in a race car, and to overlay your data with the driver's makes it easier to understand what that driver's going through. I think that data logging and driver coaching have probably gone hand in hand and are probably growing equally.'

Thought-provoking stuff from Nigel Greensall then, who is clearly an advocate of the KISS philosophy – keep it simple, stupid! Another exponent of that approach is race engineer Mick Kouros, who explains his approach to the use of data.

The professional race engineer

Greek-born Australian race engineer Mick Kouros has been involved in motorsport since he was 17 (in the early 1970s), and during that long career he has worked on competition cars ranging from Mini Coopers to Group A Touring Cars, Formula Vee, Formula Ford, Formula Renault Sport, Formula Atlantic, Formula 3, Formula 5000 and Formula 3000 single seaters. As a race engineer he has won championships in Australia, New Zealand, Britain and Europe, and took the F3000 Italia series (now known as the Euro F3000 series) in 1999. He currently works in the British Formula 3 Championship.

Professional race engineer Mick Kouros.

Mick is a devotee of data logging, and a firm believer in deriving benefit from solid, reliable information. To that end he persuaded the team for which he worked in Formula Renault Sport in the late 1990s to make a substantial investment in what was, for the category at the time, an expensive and highly sophisticated system from Magneti Marelli. Mick's philosophy was (and is) that the type of analysis that was made possible by this system allowed greater benefit to be derived from limited testing time, and this meant the investment was good value for money.

Now working in F3, where currently a Bosch DAS is mandatory, Mick's fundamental approach is the same as it's always been; he uses it mainly for driver analysis, to help the driver to improve his technique and speed; and, as he puts it, 'you don't need to get over-technical'. He also makes the point, as Nigel Greensall did above, that the handling is essentially driver-induced, and that the 'car has to be really bad to be the dominant factor'.

Mick's chosen channels to provide the driver with the most help are throttle,

Mick Kouros downloading data from a Formula Renault Sport single-seater back in 1997.

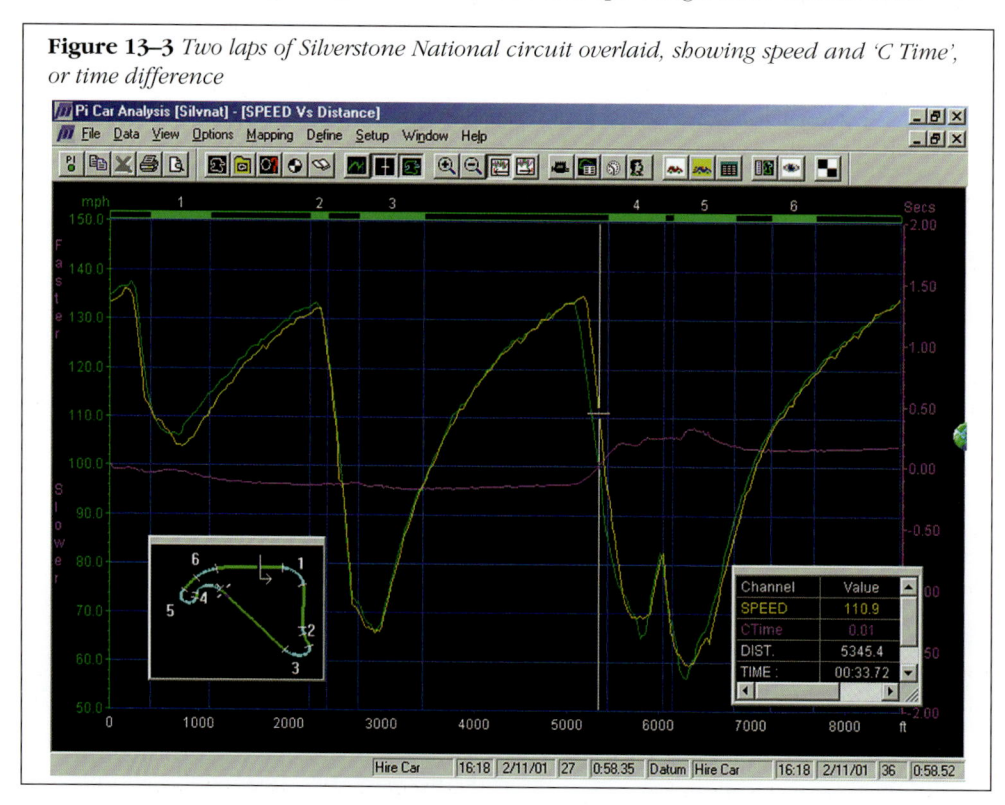

Figure 13–3 Two laps of Silverstone National circuit overlaid, showing speed and 'C Time', or time difference

speed, longitudinal G and steering angle. He also uses the time difference channel, which calculates the time difference between two selected laps, often put up by different drivers, an example of which is shown on the screen plot in Figure 13-3. From this channel it is obvious very quickly where one driver is gaining or losing time relative to another.

Mick's channels of choice are, when it comes right down to it, the ones that really matter to the driver; speed is in there because that's what it's all about and it's the easiest channel to read, while throttle, long G and steering relate directly to the three basic control mechanisms at the driver's disposal. He isn't bothered by symptoms of frequent and rapid driver corrections showing up in the data though, reckoning that the first channel he looks at is the throttle position – 'if the driver doesn't have to lift off the throttle, don't bother trying to tune out understeer or oversteer'. The corollary of that comment is that if the driver does have to come off the throttle, then maybe something does need to be done – but to car or driver?

In respect of looking at an rpm trace though, Mick's perspective perhaps reflects the high level at which he works. He reckons that provided the driver uses his shift lights properly then the rpm channel isn't particularly useful – clearly by the time they reach Mick's tender ministrations drivers are expected to be able to change gear properly! He does concede that looking at rpm can help optimise the gearing, especially in the slower corners where keeping the engine in its power band is important. Lateral G is also useful, Mick says, and he will compare lateral G in left and right-hand turns (they should be the same unless some turns are of the non-event type), and also check on the maximum lateral G achieved (compared to long G) to ensure that the driver is utilising the grip available. But he cautions that you can go too far with this, given that you will get different amounts of grip on different days at the same track.

Mick's choice of X-Y graphs was actually displayed in Chapter 11; he exploits the lateral G versus long G traction circle plot to check that maximum lateral G matches maximum long G, which he says it should on a 'wing car', that is, one that develops a lot of its grip from downforce, and the lateral G versus steering angle plot to look for signs of understeer and oversteer. Although he agrees you can spot symptoms of those imbalances just by looking at the steering trace, the relation to lateral G makes them easier to pick out.

Mick also uses damper pots and strain gauges on the suspension to help with chassis tuning. 'Even without using maths channels you can see how much movement there is, where the bumps are, how much front to rear downforce you've got which helps to maintain a stable pitch angle (rake). By using the maths channels you can calculate pitch and roll angles, but you don't actually need the maths to see them.'

However, the ability to generate maths channels can be useful: 'Over the bumps you usually need the damper velocity peaks to align with the load peaks, although this may vary if you use the dampers to fine tune oversteer or understeer.'

When starting out with a new car with which he is unfamiliar, Mick uses some static set-up logging methods to help with subsequent on-track logging. In the comfort of the workshop he logs the static strain gauge readings; he then lifts each end of the car in turn to the point where it is just unloading, and logs the strain gauge readings; and he then lifts each end further to the point where the tyres only just drag on the floor, and logs the strain gauge readings again. He is then armed with yard sticks to judge, from the data, when any of the wheels is unloading or in the air on track, and in the case of the rears, for example, this helps to deduce where wheel spin may be likely to occur. But once he has got a car into the 'tuning window' he relies on this type of logging much less.

Thorough preparation is a hallmark of Mick's approach to running a race team. This is a strategy that is also inherent amongst the teams in the world's top race series, but here yet more analytical tools supplement the gathering and use of data, as explained in the next section.

Data logging and simulations

In 2001, the CART Fedex Championship Series visited Europe for the first time in its present format, and came to two new circuits, the EuroSpeedway Lausitz in Germany and the Rockingham Motor Speedway in England. As with any visit to new circuits this presented the teams with a challenge to find what they hoped would be a competitive chassis set-up and aerodynamic package before arrival. But typically unpredictable, or should that be predictable Europe-in-September damp weather meant that practice at both venues was severely curtailed. This put even greater emphasis on pre-event set-up preparation. I was privileged to sit in with one of the teams, Herdez Bettenhausen, to follow their preparations for running on the 1.479-mile Rockingham track for the first time.

Herdez Bettenhausen was running a Lola B1/00 Champ Car powered by a Cosworth XF engine for Mexican driver Michel Jourdain Jr, who became the youngest driver ever to compete in a CART race at Long Beach in 1996, and scored a Champ Car career-best third place in the Herdez Lola at the Michigan 500 in July 2001.

Losing valuable running time in practice because of the weather and track conditions (the track refused to dry as quickly as was hoped) put greater than usual onus on the interpretation of some basic data logged in earlier testing, and the set-up derived from pre-race simulations. Alex Barron and Johnny Herbert had done some runnning around the Rockingham track a few weeks previously, and CART had supplied all the teams with some of the data logged during that running. They also supplied the race engineer's report on the test car's set-up (it was a 2000 Lola B2K/00 chassis fitted with a 2001 variant Cosworth XF engine) covering ambient and track conditions, tyre sets and aerodynamic configuration to each team. All the teams were given access to Alex Barron to discuss his thoughts on the track.

Herdez Bettenhausen's chief engineer, Scotsman Tom Brown described their preparations four weeks before the Rockingham event. 'The selection of data from testing that we were given has allowed us to make comparisons with other tracks. The information is an enormous help. It's given us real-time data to work with. The initial thinking is that Rockingham will be similar in some respects to Nazareth (the 0.946 mile Pennsylvania tri-oval). Once the aerodynamic package has been decided upon (by CART) and we have all the aero data we need from Lola we'll be able to run simulations. We'll look at the effects of downforce, drag and weight.

'The simulations will define the number of pit stops and fuel windows. But the levels of grip we'll encounter are uncertain. We believe Rockingham to be a high grip surface. This might lead to high tyre wear, so early on the first day we hope to run a set of tyres as long as possible to assess the wear and see how the track comes in and the balance changes . . .'

'We think we're OK on gearing. There is a possibility we'll be shifting, though we're not sure by how much. Big gaps (between gears) can destabilise the car, whilst it's not worth shifting if the simulations indicate small gaps. The driver also has a threshold of what he'll accept in this regard too, so it's a compromise to be considered. The data also gives us a guide to the spring rates and ride heights we'll use, in reference to similar tracks and similar speeds. Cambers and toes likewise will be set this way, and in the light of experience.'

Once the teams arrived at Rockingham, Herdez Bettenhausen's electronics and data engineer Brandon Fry expanded on how the data was utilised: 'We were given two files, showing speed traces, lateral G, longitudinal G, and front and rear brake pressures. The speed trace gave us some idea of what to expect. The lateral G data provided a track map and also a guide to the grip level of the surface, though experience tells us that this improves through a meeting as rubber gets laid, so speeds were also likely to come up through the meeting. The data showed a peak speed of 204mph, but we're predicting higher speeds than that.

'We use D.A.T.A.S. RaceSim software to do simulations,' continued Brandon. 'We used the new data, but we had to make assumptions in a number of areas, for example springs and ride heights. We also had to speculate on the grip factor. The software allows you to "tweak" the local grip factor to correlate with test data, and then you can look at an overall increasing grip level. Normally you would recalibrate this through practice on the first day of a meeting.'

However, with just 90 minutes of pre-race running possible on race day morning, there was little opportunity to do other than log some data and check that against the simulations to ensure the fuel- and tyre-stop strategy was about right. The team uses the high-end Pi Sigma data acquisition system, logging 30–40 chassis channels including all the basic ones like speed, throttle, steering, lateral and longitudinal G, plus damper displacements, pushrod load cells, tyre pressures (using the Pi HPTMS system) and so forth. A microwave telemetry link transmits the data back to the pits so Brandon and Tom can watch what's going on, as it happens. Then there are the engine parameters, although that's dealt with via the engine control unit, supplied and jealously guarded by engine supplier Cosworth!

Amidst a general flurry of activity, data is being downloaded, seemingly from the back of the driver's crash helmet! Now there's a thought . . .

Tom Brown subsequently related that the team didn't actually have to alter much on the car, and that it was pretty well balanced right from the start of running: 'We raised the front ride height slightly because the car was touching the track at times, and we had to compensate for that by adding a little front wing. We also changed the tyre pressures to make sure that they settled at the right rates – this was different at Rockingham to other tracks because there is almost constant lateral G, the cars are almost never running straight.

'We hit our simulations OK, although the speeds were much faster than the tests, especially the corner speeds.' In fact, Michel's best lap averaged 208mph, faster than the maximum speeds set in testing and more than one driver stated that on a clear lap turns one, two and three were flat on the throttle, and only a part-lift was needed for turn four.

The race, which was shortened because of the now cramped schedule, was a mixed affair for Michel. On lap 50 he spun coming off turn four, triggering a five-lap 'yellow' caution period. He got the car pointing the right way at the end of the front straight having amazingly neither hit nor been hit by anyone, but he suffered a longish pit stop as the Herdez Bettenhausen crew replaced the front damper cover that became dislodged as he travelled backwards down the main straight (his lap time was reportedly only three tenths slower that time around!) as well as refuelling him. This delay dropped him to the tail of the field, and he ultimately finished in 19th place. But Michel later told me: 'The car was good right out of the box and if we had more time (in practice) to work on a couple of little changes we could have been very competitive.'

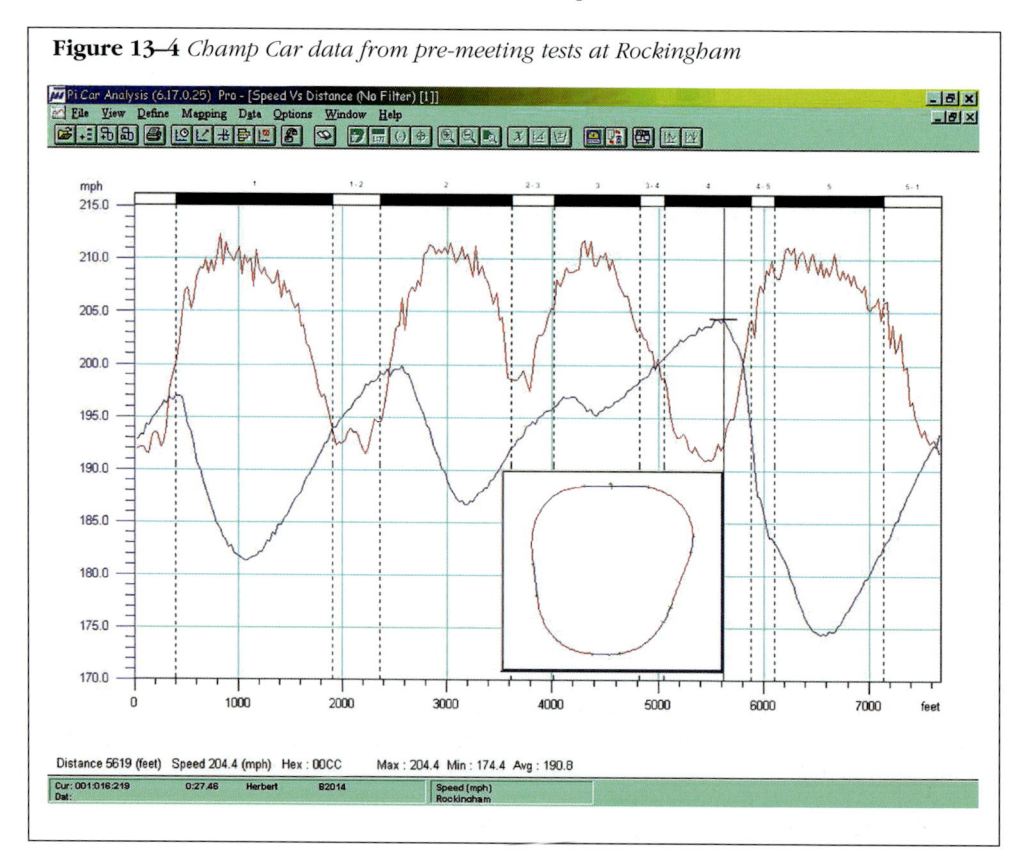

Figure 13–4 *Champ Car data from pre-meeting tests at Rockingham*

DATA LOGGING IN USE

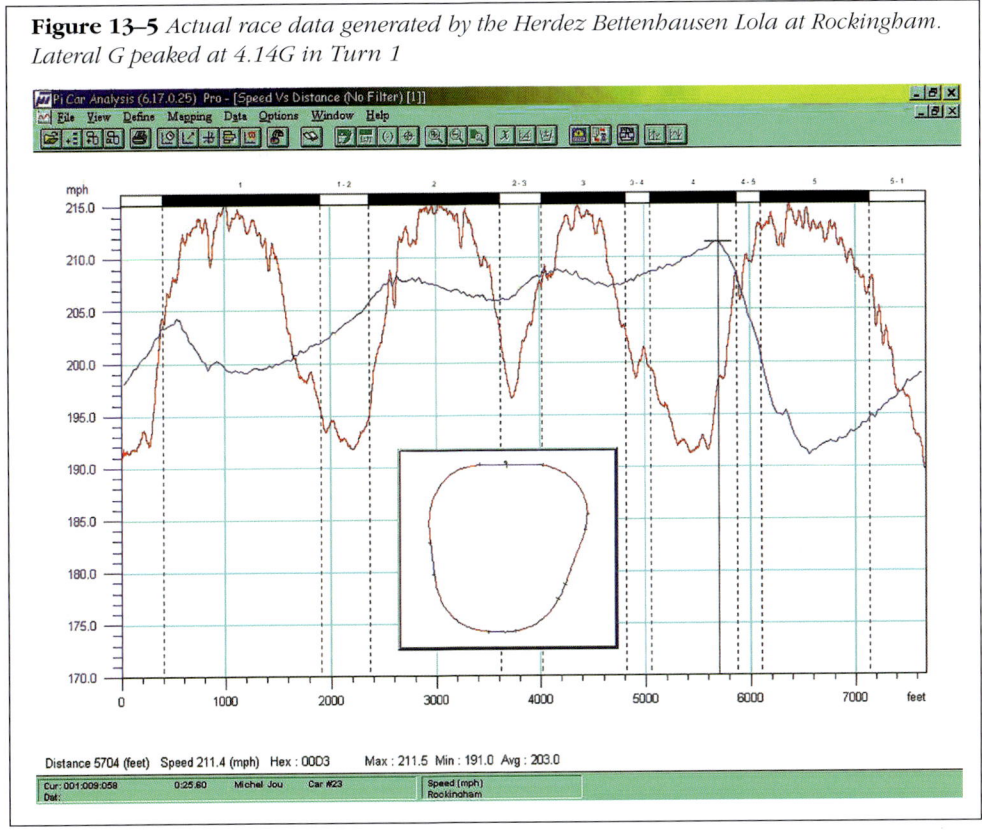

Figure 13–5 *Actual race data generated by the Herdez Bettenhausen Lola at Rockingham. Lateral G peaked at 4.14G in Turn 1*

That the car was so well balanced first time out on a new track is testimony to the value of data being available, and also to the simulations done by Herdez Bettenhausen. Also the ability of all the teams to race safely at such high speeds on a new track with so little actual running time highlights the benefit of these computer-aided preparation methods.

It's worth pausing briefly to look at Figures 13-4 and 13-5 because they give an insight into what oval driving is about. The former shows the speed and lateral G traces put up by Johnny Herbert in testing, the latter shows the same two channels (all that could be released for our delectation because of confidentiality issues) generated by Michel Jourdain Jr in the race. As you can see, the minimum speed set by Michel was around 191–192mph in turn four (17mph faster than Herbert had done on the same corner in testing), and the maximum speed was 211.4mph just before turn four. Peak lateral G readings were all just over 4G (the scale is not included in the traces), with a maximum of 4.14G in turn four, although the other turns generated nearly the same G levels. It is also evident that the lateral G declined for very little time between turns, mirroring Tom Brown's remark above, and in fact between turns two and three lateral G was always present, almost making them one long curve.

Herdez Bettenhausen is one of the more compact of the Champ Car teams, running just one driver in the series. But even the best funded teams in this most accessible of top flight race series look under-staffed compared with the front-running Formula 1 teams. Let's take a general look at how the F1 teams go data logging.

200+mph, 4+ lateral G, and a concrete wall to greet you if you lose your grip in a Champ Car

Formula 1 data logging technology

With the rapid pace of change and technological progress in F1 nowadays, no doubt this section will be out of date well before the rest of the book, and quite probably before the thing is published. But a quick and necessarily general look at what goes on makes you wonder where all the resources come from at this level of motorsport.

Take a hypothetical Grand Prix team to illustrate the resources that are employed in this area of activity. At a race weekend, Team McWillari International have with them their technical director who, amongst many other things, is in charge of all the race engineers. There are two race engineers per car, one of whom communicates with the driver, and works principally on the mechanical and aerodynamic set-up issues, while the other works solely on data analysis. In addition, for each car there is also a control systems engineer, and an electronics technician. These last-mentioned people are never usually visible, working away in the background, either tucked behind the screens at the back of the pit, or else in the race team truck behind the pits.

In terms of physical resources there will either be networked workstations or laptops (the latter are preferred at McWillari), but either way there are miles of cable connecting all the systems together. There is also an optical link from the pit garage to the pit wall crew, who have a further dozen monitors in front of them supplying information from the team and the timekeepers.

The cars themselves in F1 mostly have data logging and telemetry systems supplied currently either by Pi Research, TAG Electronics or Magneti Marelli, although a couple of teams still do their own chassis electronics and data acquisition. All the teams have their own control systems software engineers. From 2001 the teams were once more allowed to use any electronic control systems on the engines and power train, although electronically con-

Not quite as portable as your average laptop. Formula 1's answer to data logging and analysis. (Tracey Inglis)

trolled steering and braking systems are no longer permitted. But with the new engine and power train control freedoms came the ability to use any sensor as a control input, for example for differential, launch and traction control.

In terms of pure data acquisition, there are something like 65 chassis sensors on the racecars (over 100 on the test cars). In the constant search for reducing weight, the list of sensors regularly comes in for review, but doesn't ultimately shrink at all. The benefit of the data literally outweighs any penalty directly attributable to the few tens of grams that most sensors physically weigh.

A list of the logged channels includes suspension load cells, damper positions, a three-axis accelerometer (lateral, longitudinal and vertical G), wheel hub accelerometers, Pitot tube (for air speed), yaw rate (a gyroscope, though solid state devices are on their way), steering angle, wheel speeds, position and speed based on GPS (Global Positioning System, using signals from Earth-orbiting satellites) throttle position, clutch pressure, clutch position, gear change barrel position, brake wear, hydraulic system pressure, driveshaft torques, brake line pressures, ride height (up to four laser sensors), gearbox shaft speeds and torques. It's an impressive list, and the data is viewable in real time, as it happens on track, via the telemetry link back to the networked computers and monitors in the pit.

But just like most of the rest of us, who may only be logging between one and six channels, the F1 guys still have to develop ways of quickly extracting the most useful data from the most useful channels in order to diagnose problems, or suggest solutions to those problems.

Last word
Which brings us neatly to the last few bits of advice from the professionals who use data logging on a daily basis;
- Keep all your data, backed up on floppy discs, CDs, or whatever medium you favour.
- Have ready access to your key data (fastest or datum laps or runs).
- Develop methods of taking a 'quick look' at the things that matter most to you.
- Make sure you maintain, calibrate and check everything regularly.
- If something doesn't look right, investigate – you might regret it if you don't.

And on that last note, just one final anecdote from someone who will remain nameless out of sympathy. An rpm trace started to show spurious readings, with values sporadically about half what they should have been. The other channels were working OK, so the rpm problem was put in the 'fix it later' category on the basis that it didn't seem crucial. Then the flywheel starter ring, from which the rpm readings were being taken, fell off. The spurious readings were caused by the starter ring working loose . . . Happy logging.

Appendix A

First, unpack the computer!

- and other useful tips for novices

SO YOU'VE GOT your new desktop PC or laptop out of its box, and after reading the instructions you have, somehow, still managed to connect all the cables in the right places and started the thing up. If you have a desktop PC, you will have a CENTRAL PROCESSING UNIT (or CPU, either as a BASE UNIT, which usually

A central processing unit in tower format.

resides under the monitor, or a TOWER UNIT which usually sits beside the monitor or on the floor), a MONITOR, and a KEYBOARD. You will also have a device called a MOUSE. All these components will now be joined by cables. If you've bought a laptop you may not have any cables attached to anything as yet, but at some point you will have, even if it's only to recharge the battery. You probably won't have a mouse with a laptop either, although you may choose to buy and attach one later.

All being well then, if you've pushed the right button to power up your new computer and you have marvelled at all the clicks, whirrings, and changing screen images that went on whilst your system 'booted up', or started as you might prefer to say, you are now staring at a screen with little coloured graphical images on.

A monitor display, with icons scattered about the 'desktop'.

Before going any further, here's some basic jargon so you can sound fluent in computer-speak right from day one! Sadly it will be impossible to avoid using jargon, so you might as well start getting used to it. Before long you'll be using it unconsciously, if not actually in your sleep. But it will also help to understand explanations, here or elsewhere, of how to make things happen on the computer.

Some computer jargon

CPU (Central Processing Unit): the brains of your computer. Comes in tower format (stands next to your monitor or on the floor) or Base Unit format (sits underneath your monitor).

VDU (Visual Display Unit): also known as **MONITOR** or screen.

MOUSE: a device that connects to the **CPU**, usually by cable, which is used to communicate instructions to the computer. If you have a laptop, you will have some other form of 'pointing device', either below or within the keyboard, that has the same functions as a mouse.

CURSOR or **POINTER**: the little arrow on the screen that moves as you move the mouse around on a flat surface (or move your finger around on your laptop's pointing device). The cursor can also appear as a little cross, or a vertical line, or an I-shape depending on what you are doing at the time. It allows you to see what options or instructions you are selecting, and by moving it around you can select the option or instruction of your choice.

DESKTOP: the picture you see when you start your computer.

ICON: little graphical images on your desktop, sometimes also called shortcuts.

WINDOW: When you start a piece of software or a program, it will appear in its own 'window' on screen. You can have several windows open on your desktop at once, running several different programs. Windows® software took its name from this concept. Note: The more windows you have open at once, the slower your computer will do tasks.

WALLPAPER: the background picture on your screen desktop. You can select a picture of your own choice if you wish.

SCREENSAVER: Cuts in automatically with a predominately black picture, originally designed to save your monitor from 'burning out' by having too much on the screen for too long. Has now become more of a fashion statement that probably does not 'save' your screen at all. Go for the standard, black-background screen savers, they might actually do some good. To get back to a working screen once the screensaver has appeared, just tap any key on the keyboard, or move the mouse or pointing device.

FLOPPY DISK: a small, square magnetic storage disk contained in a rigid plastic housing (hence, they're not floppy at all but they used to be, a long time ago!). Programs or data can be stored on and retrieved from floppy disks.

CD: Like music CDs, CDs for computing can store masses of information (programs or data). It is possible to copy data on to the right type of CD with an appropriate 'CD Writer' unit.

How it all works (roughly . . .)

The CPU is the brain of your computer, and it's where all the jobs you tell the computer to do are carried out. The screen lets you know what the computer is doing, or has done.

The keyboard is one means of communicating with the computer. The mouse, or pointing device, is the other, and is perhaps the most 'foreign' concept in computing to get used to initially. It requires hand to eye co-ordination that becomes second nature very quickly, but it is strange to start with.

The mouse nestles in your hand with the buttons underneath your fingers. Moving the mouse around on the flat surface beside your PC, preferably on a mouse mat, controls the movement of the pointer on the screen. By moving the pointer on the screen over an icon or 'button' on the screen, and 'clicking' with

APPENDIX A

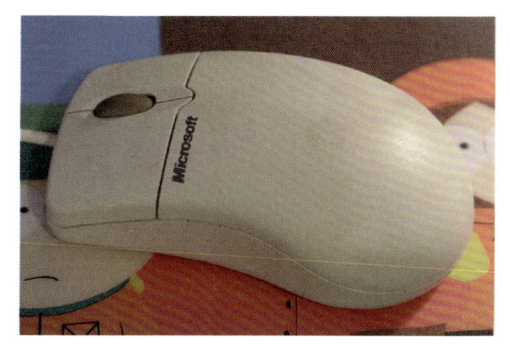

A mouse, used for selecting icons and other options.

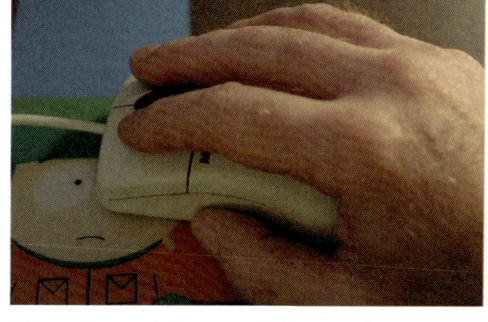

Mouse in captivity, now fully tamed.

your index finger on the left mouse button, you tell the computer to do whatever is written on the icon or button. It's called clicking because that's the noise made when you push down on the mouse buttons. Often you need to do a rapid double-click on the left mouse button to start a program running, although a single click will initiate some instructions. Right-clicking with the middle finger on the right mouse button is sometimes used to open up additional options to select.

With a laptop, the pointer is moved around by moving your finger on whatever type of pointing device you have. The equivalent of mouse-clicking is done with the aid of special buttons on the keyboard that do the same as the mouse buttons.

Most software is provided with sufficient written instructions that, with the

A laptop with mouse connected.

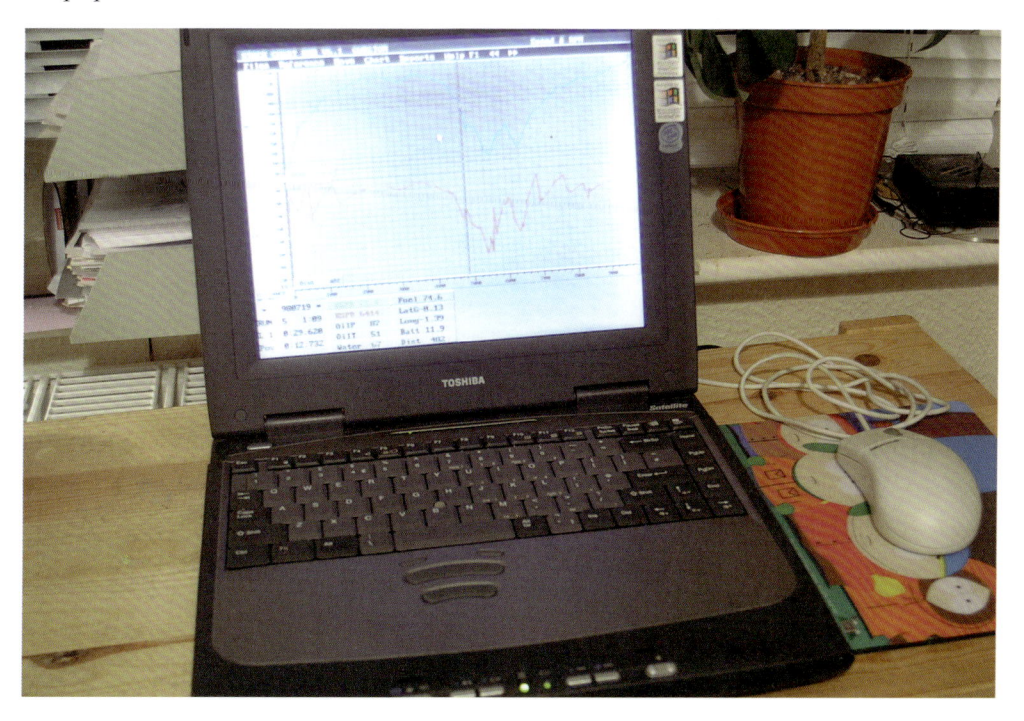

help of the above jargon-buster, you will now be able to load and run your programs. The instructions will tell you to put the floppy disk or CD provided into the appropriate drive (see your computer handbook), and then follow on-screen instructions that will involve much clicking and occasional typing.

Before you load anything on your computer though, it's a good idea to load a proprietary anti-virus program. It's so cheap compared to the havoc a virus can cause. Once that is installed you can compute with confidence. Viruses are rogue programs written by the equivalent of vandals which set out to cause inconvenience at best and actual damage at worst. They can get on to your computer system via the Internet, e-mails, disks of unknown origin and even disks from known, previously trusted sources. However, once you load anti-virus software, you will be protected automatically from these 'infections' almost all the time.

Your computer's storage system
All information/data/pictures/programs etc are stored on your computer as FILES. Files can be grouped together and stored in FOLDERS. Files and folders are organised in a family tree-like structure, which can have many branches, but all are eventually traced back to one folder, generally called the root directory. The root directory is usually called the C directory, which is also the name of the main storage medium, the HARD DISK, on your computer. So when you see reference to the 'C:\ drive', that's what is being referred to. The colon (:) and 'backslash' (\ ) are just symbols in the computing context. Files and folders are organised like any conventional filing system. The C:\ drive is analogous to the filing cabinet, and the folders and files are like their paper namesakes. The floppy disk drive is referred to as the 'A:\ drive', and the CD drive will probably be allocated another letter like D:\ or E:\, depending on the make-up of your system.

How to load DAS (and other) software
Once your PC is up and running and you want to load your data logging software, place the CD or floppy disk that it came on into the relevant drive. It may start up automatically, and if so, you may then want to invoke a 'safety procedure' recommended to me by a computing expert;

1) Exit the software by clicking on the button provided or tap the 'Esc' key on the keyboard (top left), then:

2) Double-click on your anti-virus software to start it up, and 'scan' the drive that the disk is in to ensure your software disk is free of viruses. Before all the DAS suppliers contact their lawyers to refute the suggestion that their software may be infected by a virus, let me say that in all likelihood it will be just fine. But viruses have been passed on, unwittingly and unintentionally, by the most scrupulous and responsible companies, so play safe and you won't have to confront the problem.

3) If the software disk is virus-free, then rerun the loading sequence and follow all the on-screen instructions to load the software on to your system. (If you do find a virus, follow the anti-virus software recommendations and contact your DAS supplier immediately.)

Most software is self-loading. More sound advice from my computer expert is to follow the instructions on the screen and go with the 'default settings' advised. Unless you are very sure you know what you are doing, don't change the defaults advised by the software. It will tell you if there has been a problem during loading.

Non-Windows software
If your software does not run under Windows it will be necessary to follow the supplier's specific instructions to load it on to your PC. Software that does not run under Windows runs under an earlier 'operating system called DOS. Actually, there are many people who think things were simpler in 'DOS days', and certainly the DOS-based data logging software I've used works well enough. It tends to rely

more on keyboard entries than on the mouse, which makes it better suited to laptops in a way.

Computers used to run with DOS 'in the background' with Windows providing the 'friendly' interface with the user. The latest versions of Windows at the time of writing do not have DOS in the background, but they do have a 'DOS Window' that you can access, and through which you can run DOS-based software.

To access the DOS Window, click on 'Start' at the bottom left of your screen. Scroll up to 'Programs', then scroll up to 'Accessories' on the 'pop-out menu', and then scroll to 'MS DOS Prompt' on the next pop-out menu. (See, that was jargon not as yet explained, but you followed it OK. Didn't you?! Computing's not so hard . . .) Click on this MS DOS icon, and a basically black window will open up which says 'MS DOS Prompt' across the top.

In the black window it will probably say 'C:\WINDOWS>_', and the '_' will be blinking. Type 'cd c:\' with a space between the cd and the c:\, and tap the 'return' ( ) key, also known as the 'enter' key. Now the screen message will say 'C:\>_' (this is known as the 'C prompt') and you are now running in DOS mode, and can install your DOS-based software and run it from here following the specific instructions that came from your supplier.

Starting and stopping your DAS software

Once you have your software installed, if it runs under Windows you just need to double-click on the program icon on the desktop and it will start up. To close it down again when you have finished, click on the 'X' at the top right of the window. To open a non-Windows program, go to the MS DOS window as described above, change to the 'C prompt' by typing 'cd c:\', and then type in the requisite word, as stated by the supplier in the instructions to start the program. To close it down, follow the on-screen instructions.

How to save work you have been doing into folders

To save the work you have done already, no matter what software you are using, click on the word 'File' (top left corner of your screen) and click on the 'Save' button. To save as a different file name, click on the 'Save As' button. This will bring up a 'Dialog box' that will ask you what you want to call the new file. Type in your new file name and click on the 'Save' button. The file will be saved in whatever folder the software chose, unless you specify a different one.

Backing-up, which is to say, making duplicate copies of your data files and keeping them in a safe, separate location from your computer's hard disk is something you need to do from the first time you download some data. You can back up your files on floppy disks, CDs or whatever medium you prefer and for which your computer has a compatible drive. To do this is perhaps most easily done using Windows Explorer or whatever 'file manager' your system uses. This allows you to copy a file from one place to another without affecting the original file.

Having opened Windows Explorer by double-clicking on its icon (if it's not already on your desktop you can find it under 'Accessories' as explained in the section above), you then locate the file you wish to copy by looking it up in the appropriate directory, click on the file-name or its icon to 'highlight' it (it changes colour when you have done that), and 'drag' it (by holding down the left mouse button whilst moving the mouse) to the drive where your back-up medium is, such as the 'A drive' for a floppy disk, or whatever drive letter applies to your CD writer if your PC has one. When the appropriate drive letter becomes highlighted, you then release the left mouse button to drop the file copy into its new location. (Other back up media are also available – see your PC's manual).

With Windows Explorer you can create new files and folders, move, copy, delete, lose and find anything you want! Even if you mistakenly delete files or folders do not panic. They have only been temporarily moved to the 'Recycle bin' (a little grey icon on the desktop, and it can also be found, usually at the bottom of the folder list, in Windows Explorer). To retrieve files from the bin, click on the recycle bin icon, click on the file you want to get back and click on the recover button. Hey presto, it goes back to where you deleted it from.

Computing tips

1. Always have a good anti-virus software package on your system and update it regularly. Never use files/folders or software whose source you are unsure of, and run all outside disks, files and attachments (if you use e-mail) through your anti-virus software before opening them.

2. Viruses are inevitable, but though highly inconvenient they are not the end of the world. You are more likely to encounter viruses if you go 'on-line' to use e-mail or the Internet, but they can be transmitted on disks.

3. Always save what you are working on every few minutes, and back up all your files and folders on a daily/weekly/monthly basis, depending on how frequently you use them. You will at some stage need them to get you up and running again, certainly if a virus has caused a loss of data for whatever reason.

4. Never delete a file or folder if you don't know what it does. It might be crucial to the running of your machine.

5. If your system 'locks up' during use, and will not respond to any inputs from mouse, keyboard or sledge hammer (that last one was a joke) press the Control, Alt and Delete keys all together once. Wait 15 or 20 seconds to see if you get any helpful on-screen messages that let you close down any troublesome programs. If you don't, press Ctrl/Alt/Del once more. This will shut your system down and restart it. Anything you were working on will have been lost. That's why you need to save whatever you are working on every few minutes, then you'll have a saved file to open and work on again.

6. If you spill coffee over your keyboard, disconnect it from the computer and run it under a tap to clear out the coffee (and sugar), then leave it to dry. It should be fine. If you spill coffee on a laptop, disconnect it from its power source, remove its battery, and do the same.

7. No matter how bad it looks, its not that bad. The only way to really damage a PC is by dropping it, not by using it.

8. If you're getting frustrated trying to solve a problem, stand up, stretch your limbs, get some fresh air and take a break! You'll probably solve the problem as soon as you're back at the keyboard.

8. Computing is a constant learning process, so never be afraid to ask what might seem to you to be a basic question. Somebody will know the answer.

Appendix B

Mini supplier directory

The following mini-directory of companies, whose adverts or websites indicate current involvement in data logging systems and sensors, is not exhaustive and omission from this list should not be construed as adverse comment on a company's products, but rather that they did not appear during our searches (and they should therefore consider increasing their advertising budget!). The directory is included only in the hope that it may provide a useful starting point in your own searches.

Advantage Motorsport
Tel: +00 (1) 908 284 2547
E-mail: advantage_ms@hotmail.com
Website: http://advantagemotorsports.com

AIM Sportsystems
Lake Elsinore, CA
Tel: +00 (1) 909 674 9090
Fax: +00 (1) 909 674 5699
E-mail: contact@aimsports.com
Website: www.aimsports.com

Alamo Devices Inc
E-mail: info@AlamoDevices.com

Astratch Racing Technology (Time Electronics) Ltd
Botany Industrial Estate, Tonbridge,
Kent. TN9 1RH
Contact: Jim Hersey
Tel: +44 (0)1959 561890
Fax: +44 (0)1959 561865
E-mail: info@astratech.co.uk
Website: www.astratech.co.uk

CCA Racepak
67 Plantation Road, Leighton Buzzard,
Bedfordshire. LU7 7HJ
Contact: Peter Clarke
Tel: +44 (0)1525 378938
Fax: +44 (0)1525 378938

Competition Data Systems (Europe)
123 Berecroft, Harlow, Essex.
Contact: Jason Thirley
Tel: +44 (0)1279 423272
Fax: +44 (0)1279 429632

Corsa Instruments
Ann Arbor, Michigan, USA
Tel: +00 1 734 761 1545
E-mail: info@corsa-inst.com
Website: www.corsa-inst.com

Cranfield Impact Centre Ltd
Wharley End, Cranfield,
Bedfordshire. MK43 0JR
Contact: Paul Knight
Tel: +44 (0)1234 751361
Fax: +44 (0)1234 750944
E-mail: cic@cranfield.ac.uk
Website: www.cicl.co.uk

Czech-Mate Enterprises, LLC
PO BOX 4008, St Peters,
Missouri, MO 63376-9008, USA
Tel: +00 1 636 939 1053
Fax: +00 1 636 477 1371
E-mail: tech@czech-mate.com
Website: www.czech-mate.com

D C Electronics
Unit 5E, West Station Yard, Maldon,
Essex. CM9 6TR.
Contact: David Cunliffe
Tel: +44 (0)1621 856 451
Fax: +44 (0)1621 842237

Druck Ltd (Pressure sensors)
Fir Tree Lane, Groby, Leicestershire,
LE6 0FH
Tel: +44 (0) 116 231 7100
Fax: +44 (0) 116 231 7103
E-mail: sales@druck.com

EFI Technology
Contact: Ole Buhl Racing Ltd,
Roughwood House, Highwood,
Nr Ringwood, Hants BH24 3LE
Tel: +44 (0)1425 478822
Fax: +44 (0)1425 47886
E-mail: sales@efitechnology.co.uk
Website: www.efi-technology.com

Farringdon Instruments Ltd
Unit 9, Oriel Court, Omega Park,
Alton, Hampshire. GU34 2YT
Tel: +44 (0)1420 541591
Fax: +44 (0)1420 587212
E-mail: sales@farringdoninstruments.co.uk
Website: www.laptimer.com

F1 Systems Ltd
Technical Centre, Owen Road, Diss,
Norfolk. IP22 3ER
Contact: Ian Randall
Tel: +44 (0)1379 646200
Fax: +44 (0)1379 646900
E-mail: enquiries@f1systems.com
Website: www.f1systems.com

**Jon Walker Motorsport Electronics
(UK Bosch agent)**
P.O.BOX 777, Windlesham. GU20 6XX
Contact: Jon Walker
Tel: +44 (0)7074 757767
E-mail: jon.walker@pp-uk.com

Kartonix Data Logging Systems
18 Woodman Way, Milton,
Cambridgeshire. CB4 6DS
Contact: Tony Russell
Tel: +44 (0)1223 440746
Fax: +44 (0)1223 440826

Kistler (Force, acceleration sensors)
Tel: +44 (0)1420 544477

Kulite Sensors Ltd (Pressure sensors)
Kulite House, Stroudley Road, Kingsland
Business Park, Basingstoke, Hants,
RG24 8UG
Tel: +44 (0)1256 461646
Fax: +44 (0)1256 479510
E-mail: Kim@kulite.co.uk

L&R Technologies Ltd
48 Kimbolton Close, Freshbrook,
Swindon, Wiltshire. SN5 8RE
Contact: L. Liddard
Tel: +44 (0)1793 873175
Fax: +44 (0)1793 873191

Magnetti Marelli
Magneti Marelli Holding SpA, Viale Aldo
Borletti 61/63, 20011 Corbetta (Mi), Italia
Fax: +39 0297 227570
comunicazione.immagine@
    corbetta.marelli.it
Website: www.magnetimarelli.com

MM Competition Systems Ltd
Unit 3, Old Station Business Park,
Compton, Berkshire. RG20 6NE
Contact: Doug Caffell
Tel: +44 (0)8707 444666
Fax: +44 (0)8707 444888
E-mail: mail@mmcompsys.com
Website: www.mmcompsys.com

MoTeC Australia Pty
Factory 7, 8-9 Gabrielle Court, Bayswater
North, 3153, Victoria, Australia
Tel: +61 3 9761 5050
Fax: +61 3 9761 5051
Website: www.motec.com.au

MoTeC (Europe) Ltd
Factory 1, Shenington Airfield,
Shenington, Banbury, Oxfordshire.
OX15 6NZ
Contact: Richard Bulmer
Tel: +44 (0)1295 680933
Fax: +44 (0)1295 680819
E-mail: moteceurope@msn.com
Website: www.moteceurope.co.uk

APPENDIX B

PENNY & GILES CONTROLS LTD
(Position sensors)
15 Airfield Road, Christchurch, Dorset,
BH23 3TJ
Tel: +44 (0)1202 409409
Fax: +44 (0)1202 409475
E-mail: xsales@pgcontrols.com

Performance Trends Inc
PO Box 530164, Livonia, MI 48153 USA
Tel: +00 (1) 248 473 9230
Fax: +00 (1) 248 442 7750
E-mail: feedback@performancetrends.com
Website: www.performancetrends.com

Pi Research
Brookfield Motorsports Centre,
Twentypence Road, Cottenham,
Cambridgeshire. CB4 8PS
Tel: +44 (0)1954 253600
Fax: +44 (0)1954 253601
E-mail: enquiries@piresearch.co.uk
Website: www.piresearch.com

Smart.Com UK Ltd
25 Ormsby Road, Canvey Island, Essex.
SS8 0NH
Tel: +44 (0)1268 690427
Fax: +44 (0)1268 510252
E-mail: sales@vtac.co.uk
Website: www.smart-com.co.uk

Spa Design Ltd
The Boat House, Litchfield Street,
Fazeley, Tamworth, Staffordshire.
B78 3QN
Contact: Ian Maple
Tel: +44 (0)1827 288328
Fax: +44 (0)1827 260528
E-mail: spa_uk@compuserve.com
Website: www.spa_uk.co.uk

Specialist Electronics Services Ltd
Craven Court, Stanhope Road,
Camberley, Surrey. GU15 3BS
Tel: +44 (0)1276 634 83
Fax: +44 (0)1276 63327
E-mail: info@sesltd.com
Website: www.sesltd.com

Stack Ltd
Wedgewood Road, Bicester, Oxfordshire.
OX6 7UL
Tel: +44 (0)1869 240404
Fax: +44 (0)1869 245500
E-mail: sales@stackltd.com
Website: www.stackltd.com

The Strain Gauging Co Ltd
(Strain gauges)
London Business Centre, Roenteen Road,
Basingstoke, Hants, RG24 8NG
Tel: +44 (0)1256 320666
Fax: +44 (0)1256 332332
E-mail: tsgc@btinternet.com
Website: www.strain-gauging.co.uk

TAG Electronic Systems Ltd
Unit 2, Genesis Business Park, Albert
Drive, Woking, Surrey. GU21 5RW
Tel: +44 (0)1483 719643
Fax: +44 (0)1483 750249
E-mail: sales@tagelectronics.co.uk
Website: www.tagelectronics.co.uk

Variohm Components
(Position sensors)
Williams' Barns, Tiffield Road, Towcester,
Northamptonshire, NN12 6HP
Tel: +44 (0)1327 351004   Fax: +44
(0)1327353564
E-mail: sales@variohm.com

Webcon UK Ltd
(Alpha data acquisition systems)
Dolphin Road, Sunbury, Middlesex,
TW16 7HE
Tel: +44 (0)1932 787100
Fax: +44 (0)1932 782725
E-mail: alpha@webcon.co.uk
Website: www.webcon.co.uk

2D Debus & Diebold Mebsysteme GmbH
Alte Karlsruher Str, 8-76227 Karlsruher,
Germany
Tel: +49 (0) 721 944 850
Fax: +49 (0) 721 944 8529
Website: www.2d-datarecording.com

Glossary of terms and abbreviations

Accelerometer A sensor for measuring acceleration, or the rate of change of speed.

Back-up A copy of a computer file, stored in a safe and separate place to the original, that can be used to replace the original in the event of loss or damage.

Calibration The process of getting the data acquisition system to display sensor signals in easily understood units relevant to car and driver performance, such as mph and Gs.

Channel The data from an individual sensor. Each sensor occupies one channel of the data acquisition system's capacity.

Data acquisition system (DAS) A system for gathering and temporarily storing data from the systems on a vehicle. The data is downloaded to a personal computer where it can be analysed. The data may also be transmitted to a remote receiver via telemetry, so telemetry is not the same as data acquisition.

Data logging A phrase that means the same as data acquisition but which fits the cover of this book better.

Differentiation The mathematical method of calculating the slope of a graph. The slope is related to the rate of change of the data, so for example speed (or velocity) may be found by differentiating distance (or displacement) data, and acceleration may be found by differentiating speed data.

Downloading The process of transferring data from a data logging system to a computer via a cable connection.

Downshift Changing down a gear.

EGT sensor Exhaust gas temperature sensor.

File The personal computer's equivalent of a paper file, which has a unique name given to it by the user, and which contains specific information perhaps relating to one event or one session at a track.

Filtering A mathematical method of smoothing 'spikey' graphs.

Friction coefficient A value related to grip, such as between a road surface and a race tyre. A high friction coefficient tyre represents one that offers high grip.

Hertz (Hz) The unit of measurement of frequency. One Hertz (1Hz) is one cycle per second.

Histogram Also known as a 'bar graph', a histogram is a chart consisting of rectangles, or bars, whose heights are proportional to the frequency of occurrence of data within specified ranges.

GLOSSARY OF TERMS AND ABBREVIATIONS

Lap beacon A device located at trackside which transmits a coded signal to a car to indicate the end of a run or lap in the data.

Lateral G In the context of competition car performance, the sideways acceleration generated by cornering. Often thought of as synonymous with cornering force.

LCD (dash) display Liquid crystal display, the means by which data is commonly displayed to the driver.

LVDT Linear variable differential transformer, a sensor for measuring small position changes or displacements very accurately.

Longitudinal G The fore and aft accelerations generated by accelerating and braking. Can be calculated from the rate of change of speed.

Maths channel A channel created by applying a mathematical function to a set of physical, logged data, and which is then displayed as if it was a logged channel.

Noise Irregular, nuisance signals that can accompany a logged signal, but which are not relevant to it. Noise is, strictly, electrical in nature, but what may as well be noise can arise from the unsatisfactory location or mounting of a sensor.

Overlay Displaying two or more graphs of data superimposed on each other. For example, the speed data from two different laps of a track may be overlaid on one graph for comparison purposes.

Oversteer A handling imbalance in which the rear tyres do not grip as well as they should in a turn. The rear of the car swings wide in a tail-slide. The condition is also referred to as 'loose'. Characterised by steering corrections in the opposite direction to that dictated by the direction of the turn.

Pitch Usually thought of as the nose-down movement of a car that occurs under braking.

Pitot tube A pressure sensor that measures air speed, as distinct from wheel speed.

Potentiometer A device with a variable electrical resistance.

Roll The sideways rotation of a car that occurs during cornering about an axis that passes along the length of the car.

Rotary sensor A sensor, often a potentiometer, which measures angle of rotation.

RVDT Rotary variable differential transformer, a sensor for measuring small rotary position changes or displacements very accurately.

Sampling rate The rate at which a data acquisition system logs data. Usually expressed as frequency, such as 10Hz, or 10 samples of data recorded per second (and thus at 0.1-second intervals).

Sensor A component or device which changes its electrical properties in a consistent and repeatable manner in response to a single external stimulus. Its electrical output (digital or analog) is used to measure the stimulus.

Shift point/shift rpm The rpm at which an upshift is made.

Slip angle The angle to which a tyre must be turned to the direction of travel in order to generate cornering force.

Steering angle The measurement of the rotational displacement of the steering from the straight ahead position.

Strain gauge Metal foil force-measurement devices, often bonded to the surfaces of components to be measured, whose electrical output varies with mechanical strain placed on the component.

Telemetry The transmission of logged data by radio (or similar) from a competition car to a trackside (or more remote) receiver. Data can be transmitted continuously in 'real time', or in 'burst mode', wherein 'chunks' of data, perhaps a lap's worth, are transmitted as the car passes the receiver.

Thermocouple A device whose electrical resistance changes with temperature.

Throttle angle The degree of opening of the throttle, usually expressed as percentage, where 100 per cent is fully open, 0 per cent is fully closed.

Torque A twisting moment, such as that caused by the transmission of engine output along a driveshaft.

Traction Grip provided by the driven wheels.

Understeer A handling imbalance in which the front tyres do not grip as well as they should. The car is reluctant to turn, and 'pushes' towards the outside of the turn. Characterised by excessive steering lock for the radius of turn.

Upshift Changing up a gear.

X-Y graph A graph in which two possibly related sets of data are plotted against each other. For example, an X-Y graph may be created by plotting lateral G values against the steering angles logged at the same instants.

Yaw The rotation of a car about the vertical axis passing down through its centre. Like roll, yaw occurs usually as the result of cornering.

Zooming The ability of data analysis software that allows the user to select a section of a graph and expand it to fill the viewable area of the computer screen.

Index

Accelerometers 17, 18, 22, 145
Aerodynamic drag 43, 49, 69, 105, 124, 127
Aerodynamic sensors 19, 24, 124
Air speed 9, 24, 25, 125, 145
Bar graph 33, 42
Brake disc temperatures 9, 23, 126
Brake line pressures 9, 126, 145
Brake sensors 126
Braking 9, 20, 21, 38, 40, 41, 46, 47, 50, 51, 55, 68-70, 72, 73, 79, 80, 82, 88, 89, 102, 104-110, 113, 114, 116, 126, 135, 145
Cornering behaviour 50
Damper movement 9, 11
Displacement sensors 18, 20, 124
Driver analysis 137
Driver coaching 133, 134, 137
Driveshaft torque 9, 128, 145
Electro-magnetic interference 18
Environment 17, 18
Filtering 62
Gear changing 54, 135
Gearing 10, 41-46, 131, 139, 140
G-sensors 17, 18, 22, 23
Handling 10, 41, 50, 55, 74, 82, 83, 90, 92, 100, 101, 119, 121, 123, 124, 126, 131, 135, 137
Histogram 33, 34, 42, 83
Lateral G 9, 48, 49, 55, 69, 102, 104, 106, 108, 110, 113-116, 126, 135, 139, 141
Lateral G-force 10, 15, 55, 80
Line graph 31, 32, 111
Linear movement 10, 20-22
Load cells 24, 121, 124, 125, 141, 145
Local air pressures 9

Long G 104, 105, 107-110, 113-116, 126, 139
Longitudinal G 9, 48, 49, 55, 69, 102, 104, 106, 108, 110, 113-116, 126, 135, 139, 141
Longitudinal G-force 9, 69
LVDT, or Linear Variable Differential Transformer 21
Maths channels 118, 119, 139
Microwave telemetry 141
Oversteer 23, 40, 48, 52, 55, 74, 80, 82, 83, 90-92, 96-99, 107, 118, 119, 131, 135, 139
Pitch 11, 23, 108, 121, 123, 139
Pitot tube 24, 125, 145
PiVIDS 129
Position sensors 19
Pressure sensors 19, 24, 125
Propshaft torque 9
Recording tachometers 11, 14
Ride height 9, 11, 122-124, 142, 145
Roll 11, 23, 74, 121, 123, 124, 139
Rotational movement 10
Rpm 9, 10, 13-15, 19, 20, 32, 34-38, 40-54, 56-60, 74-76, 78-82, 87-89, 105, 110-113, 116, 118, 130, 131, 133-136, 139, 146
Sampling frequencies, rates 11
Simulations 140-143
Speed 9, 10, 13, 15, 18, 19, 24, 34, 38, 40-47, 49, 54-57, 59-61, 66-75, 82, 83, 87-90, 93-101, 104, 105, 107-113, 116, 119, 121, 123-127, 132-137, 139, 141, 143, 145
Steering angle 9, 10, 15, 20, 26, 55, 85, 87, 90, 99-101, 116, 118, 119, 124, 139, 145

Strain gauges 22, 24, 121, 122, 124, 128, 139
Suspension loads 9, 121, 122
Suspension movement 11, 20, 121, 123
Telemetry 9, 18, 24, 125, 127, 141, 144, 145
Temperature sensors 18, 23, 125, 126
Throttle position 9, 10, 15, 18, 20, 42, 55, 60, 74, 77, 78, 82-84, 118, 135, 139, 145
Torque sensors 128
Track-a-Mation 66
Track map 10, 13, 39, 56, 64, 66, 69, 70, 87, 115, 118, 141
Traction circle 106, 114
Traction control 114, 145
Tyre pressures 9, 141, 142
Tyre temperatures 9
Understeer 26, 41, 52, 55, 82, 83, 89-92, 99-101, 118, 119, 124, 131, 135, 139
Variable resistance potentiometers 20, 21
Video Indexed Data System 129
Wheel speed 9, 13, 15, 19, 20, 112, 125
Wheelspin 40, 47-49, 83, 104, 112
X-Y graphs, plots 34, 111, 113-116, 118-119, 139